GRAMMAR SCHOOL ALGEBRA

BY

DAVID EUGENE SMITH, Ph.D.

Professor of Mathematics in Teachers College
Columbia University, New York

GINN & COMPANY

BOSTON · NEW YORK · CHICAGO · LONDON

PREFACE

This book is intended for use in the seventh or eighth school year, either replacing the arithmetic for a time or carried along simultaneously with it. The large majority of pupils do not enter the high school, and for those who do not this book furnishes such algebra as is necessary for the intelligent reading of formulas and the solution of equations found in elementary industrial manuals. Those who continue their school work will find the subject treated in such way as to stimulate an interest in their later work, and will meet no obsolete forms that must be unlearned before proceeding.

In sequence of topics the author has continued the plan adopted in his arithmetics, that of recognizing the value of the various courses of study in use in different parts of the country. Modern curricula no longer sanction for the grammar school the plan of treating each topic but once. On the contrary, they suggest the repetition of the most important portions of algebra, although favoring a somewhat exhaustive treatment of each subject whenever it is under discussion. Of the three chapters of this book, the second covers some of the ground of the first, and the third reviews some of the topics treated in the second. The first two chapters furnish sufficient work for schools that devote part of a year to algebra and part to arithmetic. The third chapter may be used if a full year is given to the subject.

The work seeks to interest the pupil in the subject at once by showing him its utilities. The formula which the artisan meets in his trade journals and the equation which throws so much light upon business arithmetic find place in the early pages. With these applications is combined the recreation element, as seen for example in the finding of numbers which satisfy given conditions, — an element which lends much interest to mathematics.

Oral algebra, like oral arithmetic, is necessary to lead to rapidity and to an understanding of general processes. Hence enough types have been suggested to form a basis for the best of all oral work, that which comes spontaneously from the teacher and the class.

While a large number of genuine applications have been made in the domain of the pupil's present and prospective experiences, scientific and financial problems in which he has no interest have been omitted. With the applications has gone a large number of those abstract, formal problems so necessary for drill in rapid algebraic work. These "problems without content" have an interest in themselves, and give to the elementary pupil some of that pleasure which comes to the more advanced student in the discovery of positive truth in the domain of pure science.

DAVID EUGENE SMITH.

CONTENTS

CHAPTER I

THE USES OF ALGEBRA; THE OPERATIONS WITH INTEGERS AND FRACTIONS, THE EQUATION

CHAPTER II

OPERATIONS CONTINUED, FACTORING; PROPORTION; EQUATIONS

CHAPTER III

FRACTIONS CONTINUED; ROOTS; SIMULTANEOUS EQUATIONS; THE COMPLETE QUADRATIC

GRAMMAR SCHOOL ALGEBRA

CHAPTER I

THE USES OF ALGEBRA; THE OPERATIONS WITH INTEGERS AND FRACTIONS; THE EQUATION

SOME OF THE USES OF ALGEBRA

1. Numbers represented by letters.—In arithmetic we often represented numbers by letters. We learned that

If one thing costs d dollars, 5 things will cost $5 \times d$ dollars, which we write \$$5d$; and n things will cost \$$nd$.

2. How we indicate multiplication.—In algebra *the absence of a sign indicates multiplication.*

It is not so in arithmetic, for 51 means $50 + 1$; but in algebra ab means $a \times b$.

ORAL EXERCISE

1. If a rectangle is 12 ft. long and 7 ft. wide, what is its area? If it is l ft. long and w ft. wide, what is its area?

2. If a train travels at the rate of 30 mi. an hour, how far will it travel in 10 hr.? If it travels m miles an hour, how far will it travel in h hours?

3. While the hour hand of a clock passes over 5 1-min. spaces, how many does the minute hand pass over? While the hour hand passes over n spaces, how many does the minute hand pass over?

3. Rules stated by letters. — We have just seen that the area of a rectangle l long and w wide is lw. If l and w are numbers of feet, lw is the number of square feet in the rectangle; if inches, lw is the number of square inches. If a represents the area, then the statement

$$a = lw$$

is a simple rule for finding the area of a rectangle.

4. Formulas. — A rule stated in letters is called a *formula*. For example, you may have found in arithmetic that the circumference of a circle equals the radius multiplied by twice the number 3.1416, 3.1416 being represented in mathematics by the Greek letter π (pī). But it is much easier to express this rule by the formula

$$c = 2\pi r.$$

ORAL EXERCISE

1. If a triangle has a base 4 and height 6, what is the area? Suppose it has a base b and height h?

2. Given $a = \frac{1}{2}bh$, find the value of a when $b = 7$ and $h = 6$; when $b = 6$ and $h = 7$; when $b = h = 10$.

3. If an automobile has a constant velocity of 8 miles an hour, how far will it go in 3 hours? If it has a constant velocity of v miles an hour, how far will it go in t hours?

4. From Ex. 3, what meaning do you get from the statement $d = vt$? (Think of d as standing for distance.) What is the value of d when $v = 15$, $t = \frac{1}{3}$?

5. If 5 men can do a piece of work in 8 days, how long will it take 4 men, working at the same rate? If m men can do it in d days, how long will it take x men? Read from the formula a rule for solving all such examples.

6. If $c = 2\pi r$, find the value of c when $r = 5$, π having the value stated in § 4.

7. If n is an integer, does $2n$ represent an even number or an odd one? What kind of a number does $2n + 1$ represent? What is the value of each if $n = 5$?

8. Represent a number divisible by 2 (see Ex. 7); not divisible by 2; divisible by 3; divisible by 5.

9. If a man saves $5 a week, what will he save in 8 weeks? If he saves d dollars a week, what will he save in w weeks?

10. If n represents any number, how shall we represent twice the number? 4% of the number? the number plus 5% of itself? the number minus 10% of itself?

11. What is the area of a parallelogram whose base is 6 in. and height 4 in.? Read from the formula $a = bh$ the rule for finding the area of a parallelogram of any given base b and any given height h.

12. What is the interest on $200 for 2 years at 5%? Read from the formula $i = prt$ the rule for finding the interest on any principal p, at any rate r, for any time t expressed in years.

13. What is the volume of a box 8 in. long, 5 in. wide, and 3 in. deep? Read from the formula $v = lbt$ the rule for finding the volume of a rectangular solid, given the length, breadth, and thickness.

14. A man rows at the rate of 5 mi. an hour in still water. How far can he go downstream in an hour, the stream flowing 3 mi. an hour? If he rows r miles an hour and the stream flows s miles an hour, how far will he go? Read from the formula $d = r - s$ a rule for finding the distance he will be able to row upstream.

ORAL EXERCISE

1. How many 1-lb. weights will just balance 32 oz. of sugar in these scales?

2. Suppose I take half as much sugar, what about the weights? Suppose I add 8 oz. to the sugar? Suppose I take away 8 oz. of sugar? Suppose I double it?

3. If the weights on both sides of the scales just balance, what can we say about the balance if we multiply them by the same number? divide them by the same number? add equal weights to both sides? subtract equal weights from both sides?

4. Suppose I have equal sums of money in two banks, and wish to keep equal sums there. If I put $10 more in one bank, what else must I do to keep the equality? Suppose I take $\$a$ from one bank? Suppose I double the amount in one bank? Suppose I take out half of the amount in one bank?

5. The equation. — An expression of equality between two quantities is called an *equation.*

For example, $x + 5 = 7$. Here we see that $x = 2$.

6. The principles of the equation. — We have found above that the following principles are true:

1. *If equals are added to equals, the results are equal.*
2. *If equals are subtracted from equals, the results are equal.*
3. *If equals are multiplied by equals, the results are equal.*
4. *If equals are divided by equals, the results are equal.*

ORAL EXERCISE

1. What must be added to 3 to make 8? to $8 - 5$ to make 8? to $x - 5$ to make x?

2. If $x - 5 = 20$, what must be added to these equals to give the value of x? What is the value of x?

3. If $x + 5 = 30$, what must be subtracted from these equals to give the value of x? What is the value of x?

4. If $\frac{1}{3}x = 7$, what does x equal? By what did you multiply these equals to find the value of x?

5. If $5x = 15$, what does x equal? By what did you divide these equals to find the value of x?

7. **Illustrative problem.** — If $2a + 1 = 9$, what is the value of a?

1. If $2a + 1 = 9$,
2. Then $2a = 8$, by subtracting 1 from these equals.
3. Then $a = 4$, by dividing these equals by 2.

WRITTEN EXERCISE

Find the value of the letter in each of the equations in Exs. 1–12:

1. $\frac{x}{2} = 24$.
2. $\frac{y}{7} = 36$.
3. $4a = 40$.
4. $6m = 48$.
5. $\frac{1}{3}x = 17$.
6. $\frac{1}{8}n = 164$.
7. $83m = 415$.
8. $131x = 524$.
9. $a + 73 = 122$.
10. $n + 27 = 111$.
11. $m - 63 = 149$.
12. $x - 87 = 236$.
13. What number added to 173 equals 361?
14. What number is that of which $\frac{1}{3}$ equals 237?

8. Letters used in solving problems. — One of the chief uses of letters in algebra is in solving problems. A problem in arithmetic can often be more clearly solved in this way than by numbers alone.

For example, I am thinking of a number. When it is multiplied by 2, and 7 is added to the result, the sum is 33. What is the number?

Solution using a letter:

1. If I am thinking of n, then

$2n$ is twice the number,

$2n + 7$ is this added to 7,

and $2n + 7 = 33$, as stated in the problem.

2. Then $2n = 26$, by subtracting 7 from equals.

3. $n = 13$, by dividing these equals by 2.

Check or *Proof.* $2 \times 13 + 7 = 33$.

Solution without using letters:

Because twice the number added to 7 equals 33, therefore if 7 be taken away from 33 there will remain twice the number. Therefore 26 is twice the number. Therefore once the number is half of 26, or 13.

The solutions compared:

$2n + 7 = 33$.	33
Subtracting 7,	7
$2n = 26$.	2)26
Therefore $n = 13$.	13

9. We therefore see that the two solutions are the same, but that the letters make the reasoning clearer. Therefore

1. *Write a letter for the number sought.*
2. *Use this letter in the statement of the problem.*
3. *This will give an equation, as shown above.*
4. *Solve this equation.*

ORAL EXERCISE

I am thinking of a number. When it is multiplied by 3, and 7 is added to the product, the result is 40. What is the number?

1. By what letter do you wish to represent the number?

2. Then how shall we represent 3 times the number?

3. How shall we represent this added to 7?

4. How shall we express this sum as equal to 40?

5. What shall we subtract from these equals to leave $3n$ on one side of the equation?

6. What shall we do to these equals so as to leave n alone on one side of the equation?

7. How shall we check or prove that our result is correct?

As the answers to the above are given, the equations should be written on the board.

WRITTEN EXERCISE

1. I am thinking of a number. When it is multiplied by 7, and 7 is added to the product, the result is 7 times 7. What is the number? ($7n + 7 =$ how many?)

2. I am thinking of a number such that twice the number and 3 times the number together equal 35. What is the number? ($2n + 3n =$ how many?)

3. If to 3 times a certain number I add 13, the sum is 43. What is the number? ($3n +$ how many $= 43$?)

4. If to $2\frac{1}{2}$ times a certain number I add $9\frac{1}{2}$, the sum is 17. What is the number? ($2\frac{1}{2}n + 9\frac{1}{2} =$ how many?)

5. If I add $2\frac{1}{2}$ times a certain number to $3\frac{1}{4}$ times the same number, the sum is 46. What is the number?

6. If from 69 I subtract 2, the result is 17 more than 5 times a certain number. Required the number.

ORAL EXERCISE

1. I am thinking of a number. If it is multiplied by 3, and 1 is added to the product, the result is 7. What is the number?

The teacher or pupil should write the work on the board. Pupils soon come to visualize work like $3n + 1 = 7$, $3n = 6$, $n = 2$.

2. If a certain number is multiplied by 5, and 4 is added to the product, the result is 19. What is the number?

3. If a certain number is multiplied by 7, and 4 is added to the product, the result is 25. What is the number?

4. If to twice a certain number I add 3 times the same number, I have how many times that number? If this sum is 35, what is the number?

WRITTEN EXERCISE

1. Ralph's father is 39 years old, and this is 3 times Ralph's age. How old is Ralph? ($3n = 39$.)

2. Rob is a year more than 3 times as old as his sister. He is 13 years old. How old is his sister? ($3s + 1 = ?$)

3. Tom has 10 less than twice as many marbles as Frank. Tom has 20. How many has Frank?

4. Jennie's mother is 6 years more than twice as old as Jennie. Her mother is 34 years old. How old is Jennie?

5. Our club played with the Crescents. Our score was 2 more than twice theirs. Ours was 14. What was theirs?

6. Our room is 33 ft. long, and this is 1 ft. more than twice the width. How wide is the room? ($2w + 1 = ?$)

7. There are 128 girls in our school, and this number is 2 more than 6 times the number of boys in this class. How many boys are there in this class? ($6b + 2 = ?$)

10. The use of x. — While we may represent a number by the initial letter *n*, or a number of dollars by *d*, or a number of marbles by *m*, or by any other letters, *it is customary to represent by the letter* x *a number which is to be found.*

Teachers will naturally encourage pupils to use other letters occasionally, particularly initial letters, where they add to clearness. But in general the convention of algebra should be recognized.

ORAL EXERCISE

Find the value of x *in the following:*

1. $2x = 20$.
2. $5x = 60$.
3. $7x = 84$.
4. $12x = 108$.
5. $25x = 125$.
6. $125x = 250$.
7. $3x + 5 = 26$. (What does $3x$ equal? What does x equal?)
8. $2x + 3 = 19$.
9. $5x + 1 = 41$.
10. $6x + 9 = 15$.
11. $7x + 3 = 17$.
12. $8x + 4 = 60$.
13. $9x + 10 = 109$.

WRITTEN EXERCISE

Find the value of x *in the following:*

1. $12x + 3 = 135$.
2. $11x + 3 = 135$.
3. $10x + 2 = 162$.
4. $10x + 12 = 162$.
5. $23x + 9 = 170$.
6. $31x + 12 = 260$.
7. $19x + 7 = 140$.
8. $17x + 11 = 300$.
9. $42x + 4 = 130$.
10. $51x + 45 = 300$.
11. $10 + 65x = 140$.
12. $20 + 30x = 470$.

Make up problems to fit the following, writing them out like those on page 8; *then solve:*

13. $3x + 4 = 16$.
14. $2x + 8 = 16$.
15. $5x + 16 = 56$.
16. $2x + 3x = 265$.

11. Equations involving subtraction. — If from 3 times the number of which I am thinking I subtract 7, the remainder is 29. What is the number?

1. I am thinking of a number which I call x.
2. Then $3x - 7 = 29$.
3. Then $3x = 36$, by adding 7 to these equals.
4. And $x = 12$, by dividing these equals by 3.

Check. $3 \times 12 - 7 = 29$.

ORAL EXERCISE

1. What number less 2 equals 6?

2. If from a certain number I subtract 5, the result is 45. What is the number?

3. If from 4 times a certain number I subtract 4, the result is 20. What is the number?

Such problems, like all oral drill, have their greatest value when given extempore by the teacher. Rapidly given, with small numbers, the work is not difficult, and it is interesting. A *little* daily drill of this kind soon makes pupils very ready in equation work.

WRITTEN EXERCISE

1. If to 25 times a certain number I add 42, the result is 667. What is the number?

2. If to 125 times a certain number I add 50, the result is 300. What is the number?

3. If from 25 times a certain number I subtract 42, the result is 583. What is the number?

4. If from 75 times a certain number I subtract 25, the result is 200. What is the number?

5. If from 37 times a certain number I subtract 33, the result is 300. What is the number?

12. Equations involving per cents. — After deducting 10% from the marked price of some goods, a dealer sold them for \$13.50. What was the marked price?

1. Let x represent the *number* of dollars of marked price. (We need not, then, trouble ourselves to write the sign \$ each time, as we should if x represented merely the *price*.)

2. Then $x - .10\,x = 13.50$,

or $.90\,x = 13.50$,

because any number less $\frac{1}{10}$ of itself is $\frac{9}{10}$ of itself.

3. Therefore $x = 13.50 \div .90$, by dividing equals by .90,
$= 15$.

4. Therefore the marked price was \$15.

Check. \$15 − 10% of \$15 = \$15 − \$1.50 = \$13.50.

WRITTEN EXERCISE

1. What number less 10% of itself equals 72?

2. What number less 17% of itself equals 166?

3. Jack now weighs 84 lbs., which is 12% more than he weighed a year ago. How much did he weigh then?

4. A certain school gained 15% this year over the number last year. It now has 161 pupils. How many had it last year? ($1.15\,x = 161$.)

5. A dealer saved \$1968 this year from his store. This is 18% less than he saved last year. How much did he save last year?

6. A dealer was obliged to sell some damaged furniture at 10% less than cost. He sold it for \$85.50. How much did it cost? How much did he lose?

7. A village having a population of 2040 at the last census found that it had lost 15% from the number at the preceding census. How many had it before? How many had it lost?

ORAL EXERCISE

Find the value of x *in each of the following:*

1. $x + 20 = 32$.
2. $x - 3 = 18$.
3. $x - 17 = 47$.
4. $x - 12 = 40$.
5. $x + 14 = 54$.
6. $x - 14 = 60$.
7. $2x + 1 = 21$.
8. $2x - 1 = 49$.
9. $2x + 5 = 15$.
10. $2x - 6 = 54$.
11. $2 + 7x = 16$.
12. $3x - 5 = 10$.
13. $5 + 4x = 25$.
14. $10x - 7 = 63$.
15. $11x + 3 = 36$.
16. $11x - 3 = 30$.

13. Illustrative problem. — Find the value of x that makes $4 + 17x = 140$.

1. If $4 + 17x = 140$,
2. Then $17x = 136$, by subtracting 4 from these equals.
3. Then $x = 136 \div 17 = 8$, by dividing these equals by 17.

Check. $17 \times 8 + 4 = 136 + 4 = 140$.

WRITTEN EXERCISE

Find the value of x *in each of the following:*

1. $21x = 42$.
2. $37x = 407$.
3. $57x = 1197$.
4. $69x = 759$.
5. $1.06x = 424$.
6. $.63x = 144.90$.
7. $13x + 52 = 221$.
8. $17x - 19 = 287$.
9. $15x + 47 = 962$.
10. $19x - 27 = 182$.
11. $325 + 14x = 367$.
12. $81x - 9 = 1530$.
13. $411 + 11x = 774$.
14. $419x - 38 = 800$.
15. $231 + 23x = 392$.
16. $125x - 125 = 750$.
17. $125x + 125 = 1000$.
18. $101x - 111 = 1000$.

ORAL EXERCISE

1. At 5% a year, how much is the interest on \$200 for one year? for two years? for 7 years?

2. At r% a year, how much is the interest on \$$p$ for one year? for t years? If the rate of interest is r, and the principal is p, and the number of years is t, what is the interest?

3. At 10% discount, how much is the discount on \$50 worth of goods? At r% discount, how much is the discount on \$$p$ worth of goods?

14. Further use of formulas. — We have just seen that if i = interest, r = rate, t = time (in years), and p = the principal,

$$i = trp.$$

WRITTEN EXERCISE

1. If $i = trp$, find the value of i when $t = 6$, $r = 4\%$, $p = \$300$; when $t = 3\frac{1}{2}$, $r = 6\%$, $p = \$500$.

2. If $i = trp$, find the value of i when $t = 4$, $r = 6\%$, $p = \$50$; when $t = 2\frac{1}{12}$, $r = 4\%$, $p = \$600$.

3. If d = discount, r = rate of discount, and p = price of goods, write the formula for d, as in Ex. 3 above.

4. From the formula of Ex. 3, find the value of d if $r = 33\frac{1}{3}\%$, $p = \$270$; also if $r = 25\%$, $p = \$240$.

5. If a train goes m miles an hour, how many miles will it go in t hours, at this rate? If d = the distance, write the formula. Find the value of d if $m = 37\frac{1}{2}$, $t = 3\frac{3}{4}$.

6. If the population of this country increases r% every ten years, and is now p, how much will it increase in the next ten years? What will the population then be?

15. Unknown quantities. Roots. — The letters of an equation for which values are to be found are called *unknown quantities.* These values are called the *roots of the equation.* To *solve* an equation means to find its roots.

For example, the unknown quantity in the equation $x + 3 = 7$ is x. The root of the equation is 4.

16. How to represent known quantities. — *The first letters of the alphabet, in an equation, represent numbers supposed to be known.*

For example, if told to solve the equation $x + b = a$, we would take b from these equals, leaving $x = a - b$. Here the a and b are supposed to stand for known numbers.

17. The members of an equation. — The quantity to the left of the sign of equality is called the *first member;* that to the right, the *second member.*

18. Symbol of deduction. — The word *therefore* is so often used in algebra that it has a special symbol ($\therefore$).

Since $2x = 8$, $\therefore x = 4$.

WRITTEN EXERCISE

Solve the following equations:

1. $x + b = 2a$.
2. $x - a = 3b$.
3. $x + 7 = 142$.
4. $x - 17 = 69$.
5. $x + a = 125$.
6. $x - c = 127$.
7. $19x - 71 = 5$.
8. $82 + 9x = 190$.
9. $9x - 243 = 423$.
10. $17x - 62 = 57$.
11. $142 + 3x = 265$.
12. $23x - 86 = 98$.
13. $261 + 7x = 303$.
14. $9x + 126 = 621$.
15. $426 + 13x = 501$.
16. $111 + 11x = 232$.

WRITTEN EXERCISE

Find the value of the unknown quantity in Exs. 1–10:

1. $7x = 609.$	2. $12x = 204$ ft.
3. $13x = \$221.$	4. $x + 3a = 7b.$
5. $174z = 1218.$	6. $13.5z = 33.75.$
7. $123y = 1353.$	8. $x + 3$ ft. $= 7$ ft.
9. $x - 127 = 236.$	10. $\frac{5}{7}x = 255$ sq. ft.

11. The average population per square mile in Africa is 11 and the population is 126,654,000. How many square miles are there? ($11x =$ how many?)

12. Mt. McKinley is 20,464 ft. high, and it is 1606 ft. more than 3 times the height of Mt. Washington. How high is Mt. Washington? ($3w + 1606 =$ how many?)

13. If r represents the mean (average) annual rainfall in inches at St. Paul, $r + 33$ represents it at New Orleans. If the mean annual rainfall at New Orleans is 60.5 in., what is it at St. Paul?

14. The average balance of each savings-bank depositor in this country in a certain year was $417.21. This was $54.97 less than twice the balance of each depositor in Hungary. What was the average there?

15. The number of English-speaking people in the world is 8.7 million more than twice the number of French-speaking people. There are 111.1 million English-speaking people. How many French-speaking people are there?

16. The average amount paid by each person in the United States for the general expenses of the government in a certain year was $5.96, which was $1.04 less than 4 times the average amount each paid for our national pension fund. What was the average for pensions?

SOME OF THE TERMS USED IN ALGEBRA

19. Names of certain terms. — We have seen some of the uses of algebra and have found that it is often helpful to represent numbers by letters.

We now need to know the names of a few of the most important terms in algebra, especially those which we shall be using at once.

20. Coefficient illustrated. — In the expression $2x$, 2 is called the *coefficient* of x.

Just as 2 apples means 2 times 1 apple, or 1 apple + 1 apple,
so $2x$ means 2 times $1x$, or $1x + 1x$.

That is, in $2x$, 2 shows how many times x is taken as an addend.

Just as $\$\frac{2}{3}$ means $\frac{2}{3}$ of $1,
so $\frac{2}{3}x$ means $\frac{2}{3}$ of $1x$.

The expression x means the same as $1x$.

If $x = 5$, $2x = 2 \times 5 = 10$, and $3x = 15$.

ORAL EXERCISE

Write the values of the letters on the blackboard before asking the questions. Give other examples of the same kind.

1. If $a = 5$ and $b = 2$, tell the value of the following:
ab, $3a$, $5b$, $7ab$, $a + b$, $a - b$, $2a + b$, $2a - b$.

2. If $a = 2$ and $x = 4$, tell the value of the following:
ax, $5a$, $10x$, $8ax$, $a + x$, $x - a$, $7a + x$, $10a - 5x$.

3. If $a = 2$, $b = 3$, and $c = 4$, tell the value of the following:
abc, $a + b + c$, ab, $a + b$, cb, $c - b$, $c \div a$, $4b - 3c$.

4. If $p = 3$, $q = 7$, $r = 9$, and $s = 12$, tell the value of the following:
$p + q$, $q + r - s$, $s - r$, $r - 3p$, $q - 2p$, $s + p$, $p + r$.

ORAL EXERCISE

1. If I have $\$x$ and you have $\$x$, how many dollars have we together?

2. If I have $2x$ dollars and you have twice as much, how much have you?

3. What is the coefficient of x in each of the following? $3x$, $17x$, x (what coefficient is understood?), $\frac{2}{3}x$, $.4x$.

4. If there are 8 classes in this school, and b pupils in each class, how many pupils are there in all?

21. Coefficient defined. — A numerical factor written before a letter is called the *coefficient* of that letter.

22. Letters may be coefficients. — We sometimes speak of letters as coefficients. If there are a pupils in this class, and each one has x dollars, they all have ax dollars. Here a is called the coefficient of x.

Instead of one letter we may have several letters. Thus, in $2axy$, 2 is the coefficient of axy, and $2a$ is the coefficient of xy.

WRITTEN EXERCISE

1. If one bag of flour weighs x pounds, how much do a bags weigh? (Find the value for $a = 25$, $x = 96$.)

2. If a glass jar of milk weighs y pounds, and there are x of them in a basket, how much will the jars in a such baskets weigh? (Find the value for $a = 6$, $x = 12$, $y = 1\frac{1}{2}$.)

3. If one chair costs a dealer c dollars, and there are b such chairs in a set, and the dealer buys a sets, how much do they all cost? (Find the value for $a = 3$, $b = 6$, $c = 4$.)

4. Copy the following, writing beneath each the numerical coefficient:

$3ax$, $\frac{3}{5}xyz$, $40\%\,x$, $0.5ab$, $.652mn$, $12.5abcd$,

23. Factors. — The quantities which, when multiplied together, form a product are called the *factors* of the product.

24. Squares. — If two factors are equal, their product is called the *square* of either.

The product 2×2 may be written 2^2, called the square of 2. In the same way the product $25 \times 25 = 25^2$, or 625, is called the square of 25.

25. Powers. — The product arising from taking a quantity a certain number of times as a factor is called a *power*.

For example, the products

$2 \times 2 \times 2 = 2^3$, or 8, $\qquad 3 \times 3 \times 3 \times 3 = 3^4$, or 81,

are powers of 2 and 3.

That is, $2^5 = 32$, the fifth power of 2,
$3^4 = 81$, the fourth power of 3,
$a^6 = aaaaaa$, the sixth power of a,
$4^3 = 64$, the third power, or cube, of 4,
$6^2 = 36$, the second power, or square, of 6,

26. Exponent. — In the expression a^6, 6 is called the *exponent* of a, and indicates the power to which a is raised.

In a^6, if $a = 2$, $a^6 = 64$. If $a = 2$ and $b = 5$, $a^b = 2^5 = 32$.

ORAL EXERCISE

1. If $a = 3$, $b = 2$, tell the value of the following:

a^2, b^3, a^4, b^5, a^b, b^a, ab, ba, $2a$, $3b$.

2. If $a = 2$, $b = 3$, $x = 4$, $y = 5$, tell the value of the following:

a^b, ab, b^a, ba, x^a, a^x, ax, y^a, y^b, ay, xy.

3. In a^2 and $2a$, name the coefficient of a; the exponent of a. Tell what each indicates.

Abundant rapid oral drill, as in such examples, should be given.

27. Algebraic expression. — A collection of letters, or of letters and other number symbols, connected by any of the signs of operation (+, −, ×, ÷, etc.) is called an *algebraic expression.*

For example, $3a$ (since the absence of a sign between 3 and a indicates multiplication), $5 + 6x$, $2a^2 + 3b^2 + c$.

28. Term or monomial. — An algebraic expression containing neither the + nor the − sign of operation is called a *term* or *monomial.*

For example, the terms of $2x^2 - 3xy$ are $2x^2$ and $3xy$. The expression $4abx^2$ is a monomial.

29. Polynomial. — An algebraic expression composed of several terms or numbers connected by the sign + or − is called a *polynomial.*

30. Binomial and trinomial. — A polynomial of two terms is called a *binomial;* one of three terms is called a *trinomial.*

For example, $a + b$ is a binomial, $x - 3y^2 + 2z$ is a trinomial.

WRITTEN EXERCISE

1. If $a = 2$, $b = 5$, $c = 1$, find the value of each binomial in the following list: $a^2 + b^2$, $ab + bc$, $b^2 - 4c^2$, $a + b$.

2. With the values given in Ex. 1, find the value of each trinomial in the following list: $3a^2 + 4b^2c^2 + 3c^2$, $5a^2 + 2b^2 - 10c^2$, $a + b - c$, $6a - b + 4c$.

3. If $x = 7$, $y = 5$, $z = 2$, write a monomial, containing one or more of these letters, that shall have the value 70; 15; 60; 100; 25.

4. If $m = 2$, $n = 3$, $q = 7$, write a polynomial, containing one or more of these letters, that shall have the value 12; 15; 30; 100. (For example, $5m + n + 2 = 15$.)

SOME USES OF MONOMIALS

ORAL EXERCISE

1. If $a = 9$, $b = 8$, what is the value of $\frac{1}{2}ab$?

2. If $a = 11$, $b = 30$, what is the value of ab?

3. What is the area of a rectangle 3 in. by 4 in.? a in. by b in.? If area $= ab$, find the area when $a = 10$, $b = 30$; $a = 40$, $b = 20$.

4. What is the area of a rectangle a high and b long? How does the area of a triangle of base b and height a compare with this? What is its area?

5. If the area of a triangle is $\frac{1}{2}ab$, find the area when $a = 10$, $b = 20$. If the measures are in inches, the area is in what kind of units?

6. The volume of a box 3 in. long, 2 in. wide, and 4 in. high is how many cubic inches? What is the volume of a box l long, w wide, and h high? of a box a'' by b'' by c''?

7. If volume $= lwh$, find the volume when $l = 10$, $w = 4$, $h = 3$. If l, w, h stand for feet, how will the volume be expressed?

8. If the area of a circle equals πr^2, where $\pi = 3\frac{1}{7}$ and r stands for the number of units in the radius, find the area when $r = 1$.

9. If the circumference of a circle equals $2\pi r$, find the circumference when $r = 1$.

10. If a man saves \$$d$ a month, how much will he save in t months? How much will this be if $d = 15$ and $t = 6$? if $d = 25$ and $t = 10$?

11. If a boy earns c cents a day, how much will he earn in d days? If he spends a cents in this time, how much will he then have? Suppose $c = 20$, $d = 10$, $a = 75$?

Some Uses of Polynomials

ORAL EXERCISE

1. If we let a stand for 10, how may we represent 20? 24?

2. If $x = 3$, how shall we represent 3×10? 30? 34?

3. If $x = 7$, how shall we represent $3\frac{1}{2}$? 70? 74? 80?

4. If $x = 2$, $y = 3$, $z = 1$, how may we represent 200?

5. This room is a long, b wide, and c high. What is the total length of all its 12 edges?

6. Suppose your marks in arithmetic were x on Monday, y on Tuesday, and z on Wednesday, what is the average? Suppose $x = 9$, $y = 10$, $z = 8$?

WRITTEN EXERCISE

1. If $x = 3$, $y = 39$, what is the value of $37x + 5y$?

2. A man earns a cents an hour, and b cents an hour working overtime. What does he earn in an 8-hour day, working also 1 hour overtime?

3. Rob earns r cents an hour, Jack j cents, and Tom t cents. How much will all three earn if Rob works 5 hr., Jack 3 hr., and Tom 7 hr.?

4. A room is l ft. long and w ft. wide. In the middle is a rug r ft. square. How many square feet of the floor are not covered by the rug?

5. Texas has 32,290 sq. mi. more territory than five times the territory of Michigan. If Michigan has m sq. mi., what is the area of Texas? Suppose $m = 58{,}915$?

6. Richmond is 108 mi. north of the halfway point between New York and Savannah. If the distance from New York to Savannah is d, what is the distance to Richmond? Suppose $d = 904$ mi.?

THE NEGATIVE NUMBER

31. Numbers below zero. — The mercury in a thermometer stands at 68° to-day. If it falls 23° to-night, we say that it stands at 45°. If to-morrow it should fall 13° more, it would stand at 32°, the temperature at which water freezes. If it should then fall 32°, we would say that it stands at 0°, or at zero. When it goes below zero, say 5°, we indicate this fact by using a minus sign, 5° below zero being written – 5°.

32. Positive numbers. — The ordinary numbers with which we are familiar are called *positive numbers.*

33. Negative numbers. — Numbers on the other side of zero from positive numbers are called *negative numbers.*

Negative numbers are written with a minus sign before them. Positive numbers need have no sign to distinguish them, but when it is desired to emphasize their quality a plus sign may be used, as in $+5$, $+\frac{2}{3}$, which means the same as 5, $\frac{2}{3}$.

34. Two uses for the plus and minus signs. — Hence the signs + and – have two uses:

(1) To indicate addition and subtraction, *signs of operation.*

(2) To indicate positive and negative numbers, *signs of quality.*

35. Debt and credit. — To say that a man has $100, or + $100, means that he has that amount to his credit; to say that he has $0 means that he is just even with the world; to say that he has – $100 is only another way of saying that he is $100 in debt, that he has $100 on the wrong side of zero.

ORAL EXERCISE

1. If a man has $1000 and loses $900, how much has he? if he loses $100 more? $300 more? How is this written?

2. If the mercury in a thermometer stands at 40°, and then falls 40°, where does it then stand? If it falls 10° more, where does it then stand? How is this written?

3. If we call latitude north of the equator positive, how shall we designate south latitude? If we call latitude south of the equator positive, how shall we designate north latitude?

4. If a man weighing 170 lb. steps into the basket of a balloon pulling up just 170 lb., what is the combined weight of the two? Suppose the balloon pulls up 270 lb., what is the combined weight?

5. What is the altitude, in feet, of a point 1 mi. above the sea level? of a point 5000 feet lower? of a point 280 ft. lower still? of a point 100 ft. lower still? How shall we indicate this altitude?

6. If a man has $400 in the bank, and draws out $300, how much has he left? Suppose he then draws out $100? If allowed to draw $50 more (to "overdraw"), how shall we designate his balance?

7. If we call a certain point on the blackboard zero (0), and call distances to the right of it positive, how shall we designate distances to the left? If we call distances above it positive, how shall we designate distances below it?

8. New Orleans is in 90° west longitude, which we will call + 90°. What is the longitude of a place 80° east of it? of Greenwich, which is 90° east of New Orleans? of Paris, which is 100° east of New Orleans? How shall we indicate the longitude of Paris by using a negative number?

9. How much difference in price is there in selling a horse \$15 below cost or \$20 above cost?

10. The temperature on one January morning in Denver was $+ 8°$, and the next day it was $- 4°$. What was the average?

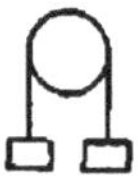

11. Weights of 7 lb. and 9 lb. hang over a pulley. How will they move? What two methods could you use to make them balance?

12. A man who was \$70 in debt paid \$50. How much was he then in debt? Suppose he earns \$50 more, how much is he then worth?

13. If there is a house for every number, how many houses would you pass in going from 21 East Washington Street to 5 West Washington Street, including both these houses?

14. Jefferson Street is 6 blocks east of Adams Street, and Monroe Street is 12 blocks west of Adams Street. Monroe Street is how many blocks west of Jefferson Street?

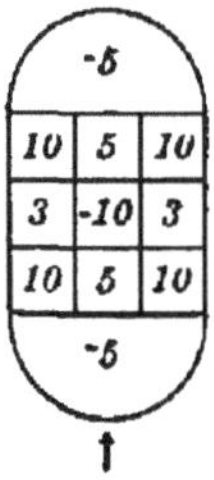

15. A game is played by throwing bean bags in the direction of the arrow. Suppose the score stands $- 5, 3, 10, 10, 0, - 10, 5, 10, 10$, how much is the total score?

16. The tide at the ocean is measured by a tide gauge. At mean (average) tide a pencil points to 0 on a scale, sliding to the right 1 space for every foot of rise, and to the left 1 space for every foot of fall. How many feet does the tide fall when the pencil moves from $+ 8$ to $- 3$?

Teachers should give a great deal of oral drill of this kind, using blackboard illustrations when necessary, until the idea of negative number as the opposite of positive number is well understood.

ORAL EXERCISE

1. If we call a force pulling upwards + 3 lb., how shall we designate an equal force pulling downwards?

2. Draw this figure on the board. Calling O zero, and the distances to the right positive, point to the distances + 3; + 5; – 2; 0; – 4; + 2½; – 3¼.

36. Curve tracing. — In this figure the successive days of the week are represented on the line OX, and the temperatures on the lines parallel to OY. The broken line shows that the temperature on one day, say at noon, was + 70°, the next day 60°, then 65°, 50°, 60°, 80°, 70°.

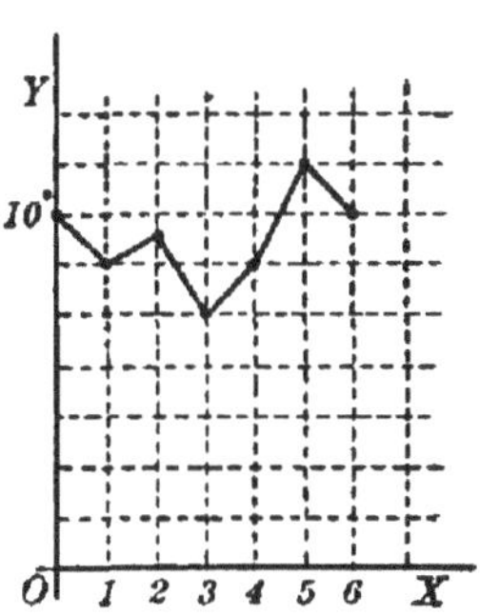

WRITTEN EXERCISE

1. On ten successive January days in Duluth the thermometer at noon registered 40°, 60°, 55°, 50°, 20°, 0°, –10°, – 15°, 10°, 30°. Trace the broken line or curve.

2. The increase in population in Nevada in four successive census years was 36,000, 20,000, – 17,000, – 3000. Represent the time differences by ¼ in., and 10,000 population by ¼ in., and trace the curve.

3. The increase in population in Virginia in eight successive census years was 200,000, 30,000, 200,000, 170,000, – 370,000, 180,000, 140,000, 200,000. Using half the scale of Ex. 2, trace the curve.

Any tables of statistics or records of temperature furnish material for curve-tracing problems.

ADDITION

ORAL EXERCISE

1. Add 2 apples + 3 apples; \$2 + \$3; $2a + 3a$.

2. Add $4x + 5x$; $7x + x$; $x + 9x$; $x + \frac{1}{2}x$; $2x + 3x$.

3. Add $10x + 15x$; $25x + x$; $5x + \frac{3}{4}x$; $10x + 0.25x$.

4. Add $2x + 3x + 5x$; $7a + 2a + a$; $3a + 3a + 3a + 3a$.

5. Add $3x + 4x + 7x$; $2n + 3n + n$; $4n + 2n + 2n + 3n$; $2b + 4b + b + b$; $8m + m + m + 5m$; $6c + 4c + 10c$.

37. Like quantities. — Quantities like $2a$, $4a$, and $\frac{1}{2}a$, that are the same except for coefficients, are called *like quantities*.

38. Adding like quantities. — We have found that *like quantities are added by adding their coefficients*, writing the letters in the sum.

For example, $3abx + 2abx + abx = (3 + 2 + 1)abx = 6abx$, where the parentheses show that the numbers inclosed are taken together.

$$\begin{array}{r} 3\,abx \\ 2\,abx \\ 1\,abx \\ \hline 6\,abx \end{array}$$

The sums of the following should be studied:

$$\begin{array}{rrrrrr} 2a & \$2 & \frac{2}{3} & 2\,abm & 2 \text{ sheep} & 2s \\ 4a & 4 & \frac{4}{3} & 4\,abm & 4 \text{ sheep} & 4s \\ 6a & 6 & \frac{6}{3} & 6\,abm & 6 \text{ sheep} & 6s \\ \hline 12a & \$12 & \frac{12}{3} = 4 & 12\,abm & 12 \text{ sheep} & 12s \end{array}$$

WRITTEN EXERCISE

1. $7a + 6a + 23a + 41a$; $19x + 27x + 69x$.

2. $17x + 32x + x + 123x$; $15a + 49a + 77a$.

3. $15ab + 16ab + ab + 23ab$; $18x^2 + 97x^2 + 39x^2$.

4. If $a = 2$, $b = 3$, find the value of $16a + 32b$. (Find the value of each separately, and then add.)

ORAL EXERCISE

Add the quantities in Exs. 1–9, reading the columns rapidly, as you read a word:

1.
$$\begin{array}{l} 7\text{ ft. }3\text{ in.} \\ 2\text{ ft. }4\text{ in.} \\ \hline \end{array}$$

2.
$$\begin{array}{l} 7\text{ sq. ft.} + 3\text{ sq. in.} \\ 2\text{ sq. ft.} + 4\text{ sq. in.} \\ \hline \end{array}$$

3.
$$\begin{array}{l} 7f + 3i \\ 2f + 4i \\ \hline \end{array}$$

4.
$$\begin{array}{l} 5\text{ gal. }2\text{ qt. }1\text{ pt.} \\ 9\text{ gal. }1\text{ qt.} \\ \hline \end{array}$$

5.
$$\begin{array}{rcrcr} 5g & + & 2q & + & 1p \\ 9g & + & 1q & & \\ \hline \end{array}$$

6.
$$\begin{array}{rcrcr} 5x & + & 2y & + & z \\ 9x & + & y & & \\ \hline \end{array}$$

7.
$$\begin{array}{rcrcr} 4a & + & 2b & + & 3c \\ 2a & + & 3b & + & 5c \\ 6a & + & 4b & + & 2c \\ \hline \end{array}$$

8.
$$\begin{array}{rcrcr} 8x & + & y & & \\ 3x & & & + & z^2 \\ & & 2y & + & 4z^2 \\ \hline \end{array}$$

9.
$$\begin{array}{rcrcr} 2w & + & 3x & + & 4y \\ 2w & + & 3x & + & 4y \\ 2w & + & 3x & + & 4y \\ \hline \end{array}$$

10. In adding algebraic quantities, as in Ex. 8, how do you arrange the like terms? Then how do you proceed?

WRITTEN EXERCISE

Add in Exs. 1–6:

1.
$$\begin{array}{rcr} 12a & + & 4b \\ 19a & + & b \\ 36a & + & 39b \\ \hline \end{array}$$

2.
$$\begin{array}{rcr} 17x & + & 31y \\ 92x & + & 73y \\ 41x & + & 9y \\ \hline \end{array}$$

3.
$$\begin{array}{rcr} 63x^2 & + & 19y \\ x^2 & + & 73y \\ 91x^2 & + & 80y \\ \hline \end{array}$$

4.
$$\begin{array}{rcrcr} 2a & + & 3b & + & c \\ 4a & + & 9b & + & 2c \\ 9a & + & 8b & + & 7c \\ \hline \end{array}$$

5.
$$\begin{array}{rcrcr} 16a & + & 5b & + & 7c \\ 3a & & & + & 9c \\ 5a & + & 10b & & \\ \hline \end{array}$$

6.
$$\begin{array}{rcrcr} a & + & b & + & c \\ 3a & + & 3b & + & 3c \\ 5a & + & 5b & + & 5c \\ \hline \end{array}$$

Add in Exs. 7–9; then find the value of each addend and of the sum, letting a = 2, b = 3, c = 5, x = 1:

7.
$$\begin{array}{rcrcr} a & + & 2b & + & c \\ 2a & + & 3b & + & 2c \\ a & & & + & c \\ & & b & + & c \\ \hline \end{array}$$

8.
$$\begin{array}{rcrcr} a & + & b & + & x \\ 2a & & & + & x \\ & & 3b & + & x \\ 2a & + & 4b & & \\ \hline \end{array}$$

9.
$$\begin{array}{rcrcr} b & + & c & + & 2x \\ 2b & + & 3c & + & x \\ b & + & 2c & + & 3x \\ 3b & + & 2c & + & x \\ \hline \end{array}$$

Equations involving Addition of Algebraic Terms

ORAL EXERCISE

1. If twice a certain number plus 3 times that number equals 50, how many times the number is 50?

2. If $2x + 3x = 50$, how many times x is 50? Then what is the value of x?

In the following examples, first find how many times a certain number you have; then find the number, as above:

3. $x + 2x = 6$.
4. $x + 3x = 12$.
5. $x + 9x = 20$.
6. $x + 8x = 81$.
7. $x + 7x = 56$.
8. $x + 6x = 63$.
9. $x + 19x = 80$.
10. $2x + 6x = 56$.
11. $3n + 7n = 90$.
12. $2y + 13y = 45$.
13. $21z + 4z = 75$.
14. $38k - 3k = 70$.
15. $51x - x = 100$.
16. $5m + 15m = 20$.
17. $14x + 2x = 32$.
18. $x + 2x + 3x = 30$.
19. $27x - 7x = 100$.
20. $x + 9x - 2x = 80$.

21. Twice a certain number plus 5 times that number equals 49. What is the number?

22. A certain number plus twice the number plus 3 times the number equals 60. What is the number?

23. A certain number plus 7 times the number plus 12 times the number equals 100. What is the number?

39. Illustrative problem. — Find the value of x when $17x + 12x + x = 510$.

1. Since $17x + 12x + x = 510$,
2. Then $30x = 510$, by adding like terms,
3. And $x = 17$, by dividing these equals by 30.

Check. $17 \times 17 + 12 \times 17 + 17 = 289 + 204 + 17 = 510$.

WRITTEN EXERCISE

Find the value of x *in Exs. 1–14:*

1. $19x + 27x + 2 = 94$.
2. $3x - 2x + 7 = 163$.
3. $x + 2x + 3x = 102$.
4. $7x - 5x - 2 = 126$.
5. $5x - 2x + 3 = 123$.
6. $x + 2x + 7x = 300$.
7. $3 + 4x + 2x = 669$.
8. $10 + 9x - 2x = 619$.
9. $6x + 7 - 2x = 275$.
10. $x + 9x + 13x = 207$.
11. $x + 3x + 5x = 108$.
12. $9x - 6 + x + 1 = 75$.
13. $x + 9x + 8x = 126$.
14. $83x + 2x + 7x + 7 = 283$.

15. On account of change of temperature a watch gains 27 sec. each day and loses 19 sec. each night. If set right to-day, in how many days will it be a minute fast?

16. If in a class of 13 boys and 15 girls each pupil has the same number of cents, and the total amount is $2.52, how much has each? ($13n + 15n =$ what number?)

17. A famous mathematician was once asked the time, and is said to have replied, "There remains of the day twice the number of hours already passed." What time was it?

18. The length of a field is $1\frac{1}{4}$ times its width. The distance around the field is 36 rods. Required the dimensions.

If x is the width, what is the length? The distance around is how many times the length and width? In cases of this kind always draw a diagram when reading the example.

19. If you tell me to think of a certain number, then to multiply it by 3 ($= 3x$), then to add to this twice the number ($= 3x + 2x$), then to add 1 ($= 3x + 2x + 1$), and I tell you the sum is 51 (that is, $3x + 2x + 1 = 51$), how do you find the number? What is the number?

20. The Milwaukees won 83 games of baseball in one year, which was 3 less than twice the number they lost. How many did they lose?

21. One of the best baseball records is that of Wagner, who was 512 times at the bat in one year. This number is 27 more than 5 times the number of runs he made. How many runs did he make?

22. A man bought a suit of clothes and a hat for $36. The clothes cost 8 times as much as the hat. How much did he pay for each?

If x equals the number of dollars paid for the hat, what represents the amount paid for the clothes? for both? What is the equation?

23. A man paid $1430 for a horse, a carriage, and an automobile. The carriage cost twice as much as the horse, and the automobile 4 times as much as the carriage. What was the cost of each?

24. A man paid $6600 for a village lot, for building a house, and for his furniture The lot cost twice as much as the furniture, and the house cost as much as the lot and furniture together. What was the cost of each?

25. At Christmas Mr. Brownson gave to each of his 3 children as many dollars as they were years old. Clara received twice as much as Helen, and Alden received as much as Clara and Helen together. All together they received $36. How much did each receive?

26. I have a farm of 175 acres, of which 15 acres are used for buildings, garden, and woodland. Of the rest, 3 times as much is devoted to oats as to wheat, twice as much to hay as to oats, and just as much to pasture as to hay. How many acres are devoted to wheat? to oats? to hay? to pasture?

ORAL EXERCISE

1. If a balloon pulling upwards 100 lb. is fastened to one pulling upwards 200 lb., what is the total upward pull?

2. If a balloon weighing -300 lb. is fastened to one weighing -200 lb., what is the total weight? What does the negative mean?

3. If a man in debt \$50 incurs a further debt of \$20, how much is his debt? That is, if a man worth $-\$50$ adds $-\$20$ to his capital, how much is he worth?

4. Add $-\$50$ and $-\$20$; -70 and -30; -6 and -12. Make a problem about -10 lb. added to -30 lb.

5. Add $-2a$, $-3a$, $-6a$; also add $-20xy$, $-30xy$, $-50xy$; also add $-6y$, $-y$, $-7y$, $-3y$, $-3y$, $-10y$.

40. Adding negative quantities. — Therefore, *to add several negative numbers, add as if they were positive, prefixing the negative sign to the sum.*

6. If a balloon pulling upwards 10 lb. is fastened to a 12-lb. weight, what is the total weight?

7. If a balloon pulling upwards 15 lb. is fastened to a 10-lb. weight, what is the total weight?

8. If in a tug of war, one class pulls 800 lb. to the right (which we will call $+800$ lb.), and another class pulls 750 lb. to the left, what is the resulting right-hand pull?

9. If a man has a capital of $-\$500$ and adds to it \$700, how much is his capital? How much is $-500+700$?

41. Adding positive and negative quantities. — Therefore, *to add a positive and a negative number, take the difference of their real values and prefix the sign of the numerically greater one.*

ORAL EXERCISE

Add the following:

1. 6 ft. – 2 in. and 4 ft. – 3 in.

2. $6x - 2y$ and $4x - 3y$; also $17x$ and $x - y$.

3. 2 lb. – 4 oz.
   3 lb. – 5 oz.

4. $\begin{array}{r} 2x - 4y \\ \underline{3x - 5y} \end{array}$

5. $\begin{array}{r} 16ab + 2c \\ \underline{-6ab + 3c} \end{array}$

42. Adding several quantities. — In adding several quantities, some positive and others negative, it is at first convenient to add all the positives, and then the negatives, then taking the difference of their real values and prefixing the sign of the numerically greater one.

$$\begin{array}{r} 3x + 2y \\ 4x - 3y \\ -x - 7y \\ -3x + 2y \\ \underline{6x - 5y} \\ 13x + 4y \\ \underline{-4x - 15y} \\ 9x - 11y \end{array}$$

Later, however, it will be found easier to read the columns, as in arithmetic. In this example we should think: "x's, 6, 3, 2, 6, 9; y's, – 5, – 3, – 10, – 13, – 11." Practice soon enables one to read the columns much more rapidly than this.

WRITTEN EXERCISE

Add in Exs. 1–26:

1. $\begin{array}{r} 32xy + 3z \\ 14xy - 2z \\ -xy + z \\ \underline{-16xy - 3z} \end{array}$

2. $\begin{array}{r} 4m^2n + 5mn^2 \\ 16m^2n - 7mn^2 \\ -9m^2n - 18mn^2 \\ \underline{23m^2n + 9mn^2} \end{array}$

3. $\begin{array}{r} 142pq + 14s^2 \\ 19pq + 37s^2 \\ \underline{-17pq - 67s^2} \end{array}$

4. $\begin{array}{r} 15a^2 - 2b^2 + c^2 \\ 16a^2 - 33b^2 - 9c^2 \\ \underline{72a^2 + 40b^2 + 6c^2} \end{array}$

5. $\begin{array}{r} 37a + 92b - 63c \\ -62a - 73b + 50c \\ \underline{25a - 19b + 13c} \end{array}$

6. $\begin{array}{r} 17ab + 3bc - 4ca \\ 6ab - 5bc + 2ca \\ \underline{-3ab + 2bc + 2ca} \end{array}$

7. $64\,a^2xy^2z + p$
$78\,a^2xy^2z + 4\,p$
$81\,a^2xy^2z + 9\,p$

8. $m^2 + m - 3$
$3\,m^2 + 4\,m$
$- 5\,m + 3$

9. $a^2 + 2\,ab + b^2$
$a^2 - 2\,ab + b^2$
$4\,a^2 - 4\,ab - 2\,b^2$

10. $4\,axy + 5\,z + w$
$3\,axy + 2\,z - w$
$7\,axy - 3\,z + w$

11. $6\,pq + 4\,qr + s$
$5\,pq - 4\,qr + 3\,s$
$9\,pq + 16\,s$

12. $156\,a - 375\,b + 48\,c$
$293\,a - 596\,b + 17\,c$
$489\,a + 783\,b - 20\,c$

13. $64\,m^2n + 92\,mn^2 + 81,\ 76\,mn^2 + 29 + 6\,m^2n.$

14. $28\,a^2b + 32\,ab^2,\ 34\,ab^2 + 29\,a^2b,\ 42\,a^2b + ab^2.$

15. $82\,pqr + 43\,qrs + 4,\ 1296\,qrs + 487 - 39\,pqr.$

16. $29\,abcd + 32\,bcde + 81\,cdef,\ 64\,bcde + 92\,cdef.$

17. $31\,a - 31\,b + 48\,ab,\ 69\,b + 52\,ab + 69\,a,\ -38\,b.$

18. $31\,a + 42\,b - 3\,c,\ 69\,a - 32\,b + 9\,c,\ -10\,b - 6\,c.$

19. $82\,p + 42\,q - 349\,r - 89\,q + 47\,p - 127\,r + 300\,r.$

20. $27\,x + 14\,y + 7\,z,\ -42\,x - 13\,y + 19\,z,\ 18\,z + 62\,y.$

21. $52m + 17n - 43p - 18p + 3m - 14n + 92p + 100m.$

22. $78\,x - 34\,y + 72z - 148y + 79x - 144x + 83z - x.$

23. $864\,x + 927\,y + 343\,z,\ 783\,y - 492\,z,\ 629\,x - 73\,y.$

24. $892a - 993b + 444c - 82d,\ 1008a - 7b + 556c - 18d.$

25. $22\,a + b - 3\,c + d - 49\,e,\ 99\,b - d - e + 3\,c + 78\,a.$

26. $427\,x^2 + 293\,xy + 829\,y^2,\ 327\,x^2 + 434\,y^2,\ 298\,x^2 + 8\,y^2.$

Add in Exs. 27–29; then find the value of each addend and of the sum when a = 1, b = 3, c = 2:

27. $2\,a + 4\,b - c$
$3\,a + 6\,b - 4\,c$
$8\,b + 9\,c$

28. $29\,a^2 + 2\,b^2$
$10\,a^2 - 3\,b^2$
$4\,a^2 + 7\,b^2$

29. $a^2 + b + c^2$
$3\,b + 4\,c^2$
$a^2 - c^2$

SUBTRACTION

ORAL EXERCISE

Subtract in Exs. 1–5:

1. \$5	2. 5 ft.	3. $5d$	4. $5f$	5. $5x$
\$2	2 ft.	$2d$	$2f$	$2x$

What must be added to the subtrahend in each of the following cases to make the minuend?

6. $9x$	7. $17x$	8. $19abx$	9. $23a + 8b$
x	$8x$	$8abx$	$3a + 2b$

In Exs. 10–12 find the differences by seeing what must be added to the subtrahend to make the minuend:

10 23 sq. ft.	11. $23f^2$	12. $18x^2 + 15y^2$
4 sq. ft.	$4f^2$	$9x^2 + 7y^2$

13. In subtracting algebraic quantities how do we arrange the like terms? Then how do we proceed?

WRITTEN EXERCISE

Subtract in Exs. 1–4, as suggested above:

1. $142x + 156y$	2. $173ax + 237by$
$97x + 68y$	$96ax + 142by$

3. $648x^2 + 273y^2$	4. $12.46abc + 2.83$
$592x^2 + 183y^2$	$6\ 23abc + 1.04$

Subtract in Exs. 5–6; then find the value of minuend, subtrahend, and remainder, letting a = 2, b = 1, c = 3:

5. $42a + 61b$	6. $32a + 15b + 20c$
$18a + 19b$	$12a + 15b + 8c$

ORAL EXERCISE

1. Ten times a number minus 3 times the number is how many times the number?

2. If 10 times a number minus 3 times the number is 42, what is the number?

3. If $10x - 2x = 16$, what is the value of x?

4. If $6x + 8x - 2x = 36$, what is the value of x?

5. Think of some number; multiply it by 6; to this product add 8 times the number; then subtract twice the number. Now tell me your result and I will tell you the number. How is it done? (Compare Ex. 4.)

6. Make up a problem like the one in Ex. 5, using other numbers. Write the statement on the board, using x or n to represent the number thought.

This presents an interesting mathematical game. Pupils may profitably propose such questions back and forth as part of the daily oral drill in solving equations.

Find the value of x *in each of the following:*

7. $9x - 3x = 54$.
8. $18x - 10x = 88$.
9. $14x - 7x = 84$.
10. $23x - 18x = 65$.
11. $47x - 2x = 135$.
12. $52x - 21x = 93$.
13. $86x - 79x = 707$.
14. $92x - 81x = 121$.

WRITTEN EXERCISE

Find the value of x *in each of the following:*

1. $x + 13x = 154$.
2. $19x - 2x = 136$.
3. $21x + 14 - 17x = 326$
4. $14x - 2x + x = 65$.
5. $3x + 47 + 19x = 223$.
6. $241 + 32 + 16x = 289$.
7. $13x + 47x + 63x = 861$.
8. $12x + 13x + 14x = 429$.

ORAL EXERCISE

1. If a man is worth − \$10, how much must be added to his capital to make him worth \$0? \$15? \$50?

2. If a man is worth − \$100, how much must be added to his capital to make him worth − \$50? − \$75? − \$25? + \$25?

3. If the thermometer registers − 20°, how many degrees must the mercury rise to register − 10°? 0°? 30°? What is the difference in degrees between − 10° and + 15°? Show this by the help of the annexed diagram.

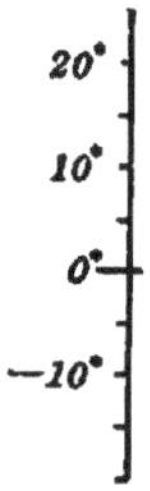

4. Draw this picture on the board. Then show what must be added to − 15 to equal 0; to equal 5; to equal 15.

5. Looking at the same picture, how much is the difference between − 10 and + 5? That is, what must be added to − 10 to equal + 5? What is the difference between − 5 and + 20?

6. State rapidly the following differences; that is, the quantity which added to the subtrahend makes the minuend:

\$15	20°	− 15°	$-10xy$	$25x^2y$
− \$10	− 15°	− 20°	$-30xy$	$-25x^2y$

7. What is the number which added to − 10 makes 0? makes − 20? makes − 5? makes − 15? (Use the above picture if necessary.)

43. How to find the difference. — Hence in algebra, as in arithmetic,

The difference is found by finding the number which added to the subtrahend makes the minuend.

44. Illustrative problems.—1. From $32x - 3y$ subtract $22x - 2y$.

What quantity added to $22x$ makes $32x$? Evidently $10x$. What quantity added to $-2y$ makes $-3y$? Evidently $-y$.

$$\begin{array}{r} 32x - 3y \\ 22x - 2y \\ \hline 10x - y \end{array}$$

2. From $-10x^2 + 7y^2$ subtract $6x^2 - 5y^2$.

What quantity added to $6x^2$ makes 0? Evidently $-6x^2$. How much more must be added to make $-10x^2$? Evidently $-10x^2$. Hence their sum is $-16x^2$.

$$\begin{array}{r} -10x^2 + 7y^2 \\ 6x^2 - 5y^2 \\ \hline -16x^2 + 12y^2 \end{array}$$

What quantity added to $-5y^2$ makes 0? Evidently $5y^2$. How much more must be added to make $7y^2$? Hence their sum is how much? What is the entire difference?

If pupils are not taught any artificial rules, but subtract in the way suggested, stopping at 0, as in the last example, if necessary for clearness, they will soon come to subtract negative quantities without difficulty. This is less confusing than to change signs and add.

WRITTEN EXERCISE

Subtract in Exs. 1–4:

1. $$\begin{array}{r} 14xy + 17yz + 5 \\ 6xy - 3yz - 2 \\ \hline \end{array}$$

2. $$\begin{array}{r} 15ab - 20bc + cd \\ 19ab - 20bc - cd \\ \hline \end{array}$$

3. $$\begin{array}{r} 17m^2 + 16n^2 + 7 \\ -17m^2 - 16n^2 - 3 \\ \hline \end{array}$$

4. $$\begin{array}{r} 62m^2 + 15n^2 - p^2 \\ 14m^2 - 20n^2 - 3p^2 \\ \hline \end{array}$$

Subtract in Exs. 5–8; then find the values of minuend, subtrahend, and difference, when a = 3, b = 1, c = 4:

5. $$\begin{array}{r} 25a - 17b + 3c \\ 32a - 9b - 4c \\ \hline \end{array}$$

6. $$\begin{array}{r} 35a^2 + 40b^2 \\ -5a^2 - 30b^2 \\ \hline \end{array}$$

7. $$\begin{array}{r} 27a + 75b + c + 7 \\ 20a + 70b - c - 3 \\ \hline \end{array}$$

8. $$\begin{array}{r} 5a + 9b^2 - 3c \\ -2a - 8b^2 + 3c \\ \hline \end{array}$$

Subtract:

9. $6abc + 4$
$9abc - 7$

10. $3m^2 - 19z^2$
$4m^2 - 17z^2$

11. $4x^2 - 3y^2$
$9x^2 - 27y^2$

12. $8p^2 + 4pq$
$7p^2 - 39pq$

13. $-1243 + x^4$
$-2987 + x^4$

14. $x^2 - 147xyz$
$x^2 - 298xyz$

15. $23a^3 + 91a^2$
$49a^3 - 37a^2$

16. $-829x^2 - 9$
$792x^2 - 8$

17. $13a^2b + 4ab^2$
$6a^2b - 2ab^2$

18. $439a - 927b$
$291a - 296b$

19. $8m^2 + 3mnp$
$6m^2 - 11mnp$

20. $-21.8x + 0.9y$
$-42.9x - 2y$

21. $-96a + b - c$
$-27a - b + c$

22. $9a^3 + 7a^2 + 2a$
$6a^3 - 4a^2 + 2a$

23. $13x^2 - 5x + 1$
$6x^2 - 5x - 1$

24. $42p^2 + 2q + r$
$-29p^2 - 3q + r$

25. $-9x - 3y - 9z$
$-7x - 9y - 8z$

26. $-92x + 3y - z$
$-79x - 3y - 9z$

27. From $89xy + 42yz + 7zw - 6$ subtract $75 - 49zw - 93xy$.

28. From $82x^3 - 48x^2 + 17x - 37$ subtract $34x - 93x^2 + 48x^3 + 43$.

Subtract, and then find the value of the minuend, subtrahend, and remainder, when a = 1, b = 1, c = 1:

29. $27a + 92b - 6c$
$-93a + 48b - 7c$

30. $a^2 + 2ab + b^2$
$3a^2 - 4ab - 7b^2$

WRITTEN EXERCISE

1. Add $x^2 + 2xy + y^2$, $x^2 - 2xy + y^2$, $8x^2 + 8y^2$.

2. Add $m^3 + 2m^2 - 3m - 1$, $m^3 - 2m^2 + m + 1$, $m^3 + 2m$.

3. Add $3abc - 4bcd + 5cda$, $-6abc + 5bcd - 5cda$.

4. From $3x^3 + 4x^2 - 5x + 6$ subtract $4x^2 - 16x + 16$.

5. From $2m^2 + 3n^2 - 2x + 3y$ subtract $n^2 - 4x + 2y$.

6. From $a + 2b - 3c + d$ subtract $a + b + c + 31d$.

7. From $6x^2 - 7xy + 4y^2$ subtract $5x^2 - 8xy + 3y^2$; then find the value of minuend, subtrahend, and remainder, if $x = 2$ and $y = 3$.

8. Add $3a^2 + 4bc - 6d$, $4a^2 - 5bc + 7d$, $3a^2 + 2bc$; then find the value of each addend and of the sum, if $a = 1$, $b = 5$, $c = 7$, $d = 1$.

Subtract, and then find the value of each minuend, subtrahend, and remainder, if a = 2, b = 5, c = 3:

9. $$\begin{array}{r} 2a^3 + 3a^2b + 2ab^2 - b^3 \\ \underline{a^3 + 4a^2b + ab^2 } \end{array}$$

10. $$\begin{array}{r} 5a^2 - 4b^2 + 3c^2 \\ \underline{2a^2 - 6b^2 - 5c^2} \end{array}$$

Add, and then find the value of each addend and sum, if x = 10, y = 1, z = 5:

11. $$\begin{array}{lll} 3x & -5y^2 & +2z \\ x & -2y^2 & \\ & 9y^2 & -4z \\ \hline \end{array}$$

12. $$\begin{array}{lll} 5x^2 & +2y^2 & -3z^2 \\ 2x^2 & & +3z^2 \\ x^2 & +y^2 & +z^2 \\ \hline \end{array}$$

13. $$\begin{array}{rrr} 2x^2 + & 4y^2 - & 3z^2 \\ 4x^2 - & 6y^2 - & 2z^2 \\ 3x^2 - & 5y^2 + & 4z^2 \\ 2x^2 & & -13z^2 \\ -5x^2 + & 2y^2 & \\ & 16y^2 - & 7z^2 \\ \hline \end{array}$$

14. $$\begin{array}{rrr} x^2 - & 3xy + & 4y^2 \\ -4x^2 + & 15xy - & 3y^2 \\ 17x^2 - & 5xy + & 52y^2 \\ 16x^2 - & 17xy - & 41y^2 \\ -9x^2 & & -13y^2 \\ & 39xy + & 17y^2 \\ \hline \end{array}$$

HOW TO USE PARENTHESES

45. Use of parentheses. — The product of $2 + 3$ by 7 is indicated by either $7\,(2 + 3)$ or $(2 + 3)\,7$, the two having the same value, 35. The sign $\times$ is evidently unnecessary.

If $a = 4$, $b = 7$, and $x = 10$, then $(a + b)x = 11 \times 10 = 110$.

The difference between 7 and $2 + 4$ is indicated by $7 - (2 + 4)$, which equals $7 - 6$, or 1.

If $a = 9$, $x = 5$, and $y = 2$, then $a - (x + y) = 9 - 7 = 2$. But $a - x + y = 9 - 5 + 2 = 4 + 2 = 6$.

The sum of $2 + 3$ and $4 + 5$ is evidently the same whether written $(2 + 3) + (4 + 5) = 5 + 9 = 14$, or $2 + 3 + 4 + 5 = 14$; so the parentheses are not necessary.

The quotient of $5 + 9$ divided by $2 + 5$ is evidently the same whether written $\frac{(5 + 9)}{(2 + 5)}$ or $\frac{5 + 9}{2 + 5}$. But

$$(5 + 9) \div (2 + 5) = 14 \div 7,$$

and this is not the same as $5 + 9 \div 2 + 5$, because

46. *It is the custom to perform multiplications and divisions before additions and subtractions* unless the parentheses show otherwise.

For example, $2 + 4 \div 2$ means $2 + \frac{4}{2} = 2 + 2 = 4$,
but $(2 + 4) \div 2$ means $6 \div 2 = 3$;
$2 + 4 \times 2$ means $2 + 8 = 10$,
but $(2 + 4) \times 2$ means $6 \times 2 = 12$.

WRITTEN EXERCISE

1. $(x + y) \div m$, where $x = 10$, $y = 15$, $m = 5$.
2. $(a + b)(c + d)$, where $a = 1$, $b = 2$, $c = 3$, $d = 4$.
3. $2 + 8 \times 2$, $(2 + 8)\,2$, $2\,(2 + 8)$, $13 + 165 \div 15$.
4. $(3 + 7) \div 2$, $4 + 6 \div 2$, $(4 + 6) \div 2$, $(7 + 19) \div (95 - 82)$.

ORAL EXERCISE

1. What must be added to $2a$ to make $5a$? Then how much is $5a - 2a$?

2. What must be added to b to make 0? to this to make a? Then what must be added to b to make a?

3. What must be added to $-b$ to make 0? to this to make a? Then how much is $a - (-b)$?

47. Subtracting negative quantities. — Let us consider the difference resulting from taking $b - c$ from a.

We may write $a + 0 + 0$ for a if we wish. Then what must be added to $-c$ to make 0? to b to make 0? to 0 to make a? Then what is the value of $a - (b - c)$?

$$\begin{array}{r} a + 0 + 0 \\ 0 + b - c \\ \hline a - b + c \end{array}$$

Because $a - (b - c) = a - b + c$, we see that

48. *If a quantity in parentheses is preceded by a negative sign, the parentheses may be removed provided the signs of the terms within are changed.*

Because $a + (b - c) = a + b - c$, we see that

49. *If a quantity in parentheses is preceded by a positive sign, the parentheses may be removed without change of sign.*

WRITTEN EXERCISE

Remove the parentheses, changing signs where necessary:

1. $a^2 + 2ab + b^2 - (a^2 - 2ab + b^2)$.
2. $m^2 - 2mn - 3n^2 + (-m^2 + 2mn + 5n^2)$.
3. $p^2 + q^2 - (p^2 - q^2) + (2p^2 - q^2) - (p^2 - 3q^2)$.
4. $a^3 - 3a^2b + (3ab^2 - b^3) - (a^3 + 3a^2b - 3ab^2 - b^3)$.
5. $a^2b + b^2c + c^2a - (a^2b - b^2c + c^2a) + (a^2b + 2b^2c - 3c^2a)$.
6. $6x^2 + (3y^2 - 4z^2) - (x^2 + z^2) + (y^2 - 4z^2) - (z^2 + 3y^2)$.

MULTIPLICATION

ORAL EXERCISE

1. Multiply by 2: 3 ft.; $4; 2 sq. ft.; 3 *f*; 4 *d*; $2x^2$.
2. Multiply by 8: 30 mi.; 31 in.; 21 sq. mi.; $40x^2$; $50ax^2$.
3. Multiply by 3: 7 ft. 2 in.; 7 sq. ft. 2 sq. in.; $7x + 2y$.
4. Multiply by 7: $a + 2b$; $9a^2 + 5b^2$; $8ab + c$; $6abx + 3by$.
5. Multiply by 6: 9 sq. ft. 2 sq. in.; $9x^2 + 2y^2$; $x^3 + 6y^3$.
6. Multiply x^2 by 3; by 25; by a; by b; by ab; by $2abc$.
7. Multiply $x + y$ by 3; by 17; by 200; by a; by $2a^2$.

50. Multiplying by monomials. — The multiplication of $x + y$ by a is indicated by $a(x + y)$, or by $(x + y)a$. Hence $a(x + y) = ax + ay$, as in the following cases:

4 ft. 2 in.	$400 + 2$	$4x + 2y$	$x + y$
5	5	5	a
20 ft. 10 in.	$2000 + 10$	$20x + 10y$	$ax + ay$

51. Therefore, *to multiply a polynomial by a monomial, multiply each term separately and add the products.*

WRITTEN EXERCISE

Multiply as indicated:

1. $23a$ by 125.
2. $421a$ by 23.
3. $43x$ by $27y$.
4. $39x$ by $46ay$.
5. $16ab$ by $15xy$.
6. $16ax$ by $15by$.
7. $25mx$ by $25ny$.
8. $25mn$ by $25xy$.
9. $abc(x + y)$; $5(a + b)$.
10. $ab(x^2 + y^2)$; $mn(a + b^2)$.
11. $x(a + b)$; $x(2a - b)$.
12. $b(p^2 + q^2)$; $5a(m + z)$.
13. $pq(a^3 + b^3)$; $b(x + y)$.
14. 3(2 da. 3 hr.); $3(2x + 3y)$.

ORAL EXERCISE

1. How many times is a taken as a factor in aa? in a^2? in aaa? in a^3? in a^4? in a^5?

2. How many times is x taken as a factor in x^2? in x^3? in $xx \times xxx$? in x^2x^3? in x^3x^4?

3. Then what is the product of b^2b^3? of b^3b^4? of m^2m^5? of m^6m? of m^4m^3? of $x^2x^3x^4$? of $x^{10}x^5$?

4. If factors are the same quantity with the same or with different exponents, what do we do with the exponents to obtain the exponent in the product?

52. Law of coefficients and exponents. — Because $2x^2 \times 3x^3 = 2xx \times 3xxx = 6x^5$, we see that if factors are the same quantities with various coefficients and exponents, we

Multiply the coefficients and add the exponents.

53. The dot as a sign of multiplication. — In multiplying in algebra we not only indicate multiplication by the absence of a sign but also by a dot, thus: $a \times b = a \cdot b = ab$, $2 \times 3 = 2 \cdot 3 = 6$.

Multiply $x^2 + y^2$ by xy. Here $xy(x^2 + y^2) = xyx^2 + xyy^2 = x^3y + xy^3$.

WRITTEN EXERCISE

Multiply as indicated in Exs. 1–10:

1. $a^2b \cdot b^2a$.

2. $ab \cdot bc \cdot ca$.

3. $xy(x + y)$.

4. $a^2bc \cdot a^3b^2 \cdot c^3$.

5. $m^2 \cdot m^3 \cdot m^2$.

6. $4x(x^2 + y^2)$.

7. $2m(m^3 + 3n^3)$.

8. $3a^2 \cdot 4a^3 \cdot 5a^2$.

9. $am \cdot 2am \cdot 3am$.

10. $82a^2(2a + 3a^2)$.

11. Multiply $2xy(x + y)$. Then let $x = 2$, $y = 3$, and find the value of each factor and of the product.

ORAL EXERCISE

1. How much more shall I weigh with two 3-lb. weights in my hands?

2. How much more shall I weigh with two – 3-lb. balloons in my hands?

3. How much more shall I weigh if I am relieved of the two 3-lb. weights?

4. How much more shall I weigh if I am relieved of the two – 3-lb. balloons?

54. The law of signs in multiplication. — The addition of two 3-lb. weights adds 6 lb. That is, $2 \cdot 3$ lb. = 6 lb.

The addition of two – 3-lb. balloons adds – 6 lb. That is, $2 \cdot -3$ lb. = – 6 lb.

The subtraction of two 3-lb. weights adds – 6 lb. That is, $-2 \cdot 3$ lb. = – 6 lb.

The subtraction of two – 3-lb balloons adds 6 lb. That is, $-2 \cdot -3$ lb. = 6 lb.

We therefore see that

55. *The product of two numbers with like signs is positive; with unlike signs, negative.*

Thus, to multiply $2x - 3y$ by $-2x$, we have $-2x \cdot -3y = 6xy$, and $-2x \cdot 2x = -4x^2$.

$$\begin{array}{r} 2x - 3y \\ -2x \\ \hline -4x^2 + 6xy \end{array}$$

WRITTEN EXERCISE

1. $3x(4x - 2y)$.
2. $-4a(a^2 - b^2)$.
3. $21xy(x - 12y)$.
4. $32a^2(a^2 + bc)$.
5. $-15a^2b(a^2 + b)$.
6. $-41x^2(x^2 - y^2)$.
7. $17a^3m(-a^3 + m)$.
8. $15xy(-3x - 7y)$.
9. $-a(-a-b-c-d-e)$.
10. $-21abc(-36a - 42b)$.
11. $-13m^2(-2m - 3n)$.
12. $a^2bc^2d(-a^2 + b - c^2 + d)$.

DIVISION

ORAL EXERCISE

1. Divide by 2: 4 ft.; \$4; $4f$; $4d$; 400; $4h$; $4x$.

2. Divide by 5: 25 sq. ft.; $25f^2$; $25x^2$; $35x^2y$; $45(x+y)$.

3. Divide by 7: 14; $2\cdot 7$; $9\cdot 7$; $x\cdot 7$; $7y^2$; $63(x^2+y^2+z^2)$.

4. Divide by x: ax; xx; x^2; $x\cdot x^2$; x^8; $x\cdot x^7$; x^6; x^9; x^7y^8.

5. Divide by xy: $3xy$; $3x\cdot xy$; $3x^2y$; $4xy^2$; $7xy\ xy$; $7x^2y^2$.

6. Divide $14a^2x$ by 14; by a^2; by x; by 7; by 2; by 28.

56. Law of coefficients and exponents. — Because

$$\frac{6x^5}{3x^3} = \frac{2\cdot \cancel{3}\cdot \cancel{x^3}\cdot x^2}{\cancel{3}\cdot \cancel{x^3}} = 2x^2,$$

we see that in division involving the same quantities with various coefficients and exponents we

Divide the coefficients and subtract the exponents.

WRITTEN EXERCISE

1. $75p^2q^5 \div 5pq$.
2. $36a^2x^2y^3 \div 4x$.
3. $32a^2b^3c^4 \div 2c^4$.
4. $39p^4q^5 \div 13p^4q$.
5. $27a^2x^5 \div 3ax^2$.
6. $45m^6n^8 \div 15n^8$.
7. $28m^3n^3 \div 14n^2$.
8. $96ax^2z^4 \div 4ax^2z^4$.
9. $144abcd \div 6ac$.
10. $1728x^6y^6z^6 \div 144x^3y^3z^3$.

11. If $16x^4y^4$ dollars are divided equally among x^2y^2 people, how much has each? Suppose $x=1$, $y=2$?

12. If $p=i+rt$ and $i=300rt$, find the value of p. (Do you know what rule of interest is expressed by $i=prt$?)

13. If $c=18\pi$ (π being the Greek letter *pi*), and if $c\div 2\pi=r$, find the value of r. (Have you learned what truth about the circle is expressed by $c=2\pi r$?)

ORAL EXERCISE

1. Divide 18 ft. 6 in. by 2; $18f + 6i$ by 2.
2. Divide 28 lb. 8 oz. by 4; $28x + 8y$ by 4.
3. Divide 10 mi. 25 rd. by 5; $10m + 25r$ by 5.

57. Dividing a polynomial by a monomial. — Divide $16a^2 + 8a + 6$ by 2.

$$2\overline{)16a^2 + 8a + 6}$$
$$8a^2 + 4a + 3$$

Dividing each term separately, we have $8a^2 + 4a + 3$.

Divide $35x^2y + 21x$ by $7x$.

$$7x\overline{)35x^2y + 21x}$$
$$5xy + 3$$

Dividing each term separately, we have $5xy + 3$.

Therefore, *to divide a polynomial by a monomial, divide each term separately and add the quotients.*

WRITTEN EXERCISE

Divide:

1. $ax^2 + ay^2$ by a.
2. $25x^2 + 35y^3$ by 5.
3. $8a + 24b$ by 4
4. $mnx^2 + mpy^2$ by m.
5. $abc^2 + abd^2$ by ab.
6. $a^3 + 4a^2 + 5a$ by a.
7. $77m + 121$ by 11.
8. $51a^3 + 187b^3$ by 17.
9. $x^2 + 2x^3 + 3x^4$ by x^2.
10. $p^2q^2r^2 + 2p^2q^2z^2$ by p^2q^2.
11. $3x^2 + 396x^2y$ by $3x^2$.
12. $4x^2y^2 + 76x^2y^3$ by $4x^2y^2$.
13. $a^4 + 3a^3 + 4a^2 + 5a$ by a.
14. $75a^3 + 25a^2 + 125a + 625$ by 25.
15. $x^4y^3z^2 + 2x^3y^2z + 3xy^2z^3 + 5xy^3z^2$ by xyz.
16. $125x^2y^2z^2 + 600xyz + 275x^3y^3z^3 + 25$ by 25.
17. $21m^5 + 35m^4n + 56m^3n^2 + 105m^2n^3$ by $7m^2$.
18. $33x^5 + 51x^4y + 54x^3y^2 + 69x^2y^3 + 111xy^4$ by $3x$.

58 The parentheses. — It should be remembered that the operations indicated within the parentheses are to be performed first.

For example, $x(x^2 + xy + y^2)$ means that the *sum* of $x^2 + xy + y^2$ is to be multiplied by x.

$14 - (8 - 6)$ means that $8 - 6$, or 2, is to be subtracted from 14. That is, $14 - (8 - 6) = 14 - 2 = 12$.

WRITTEN EXERCISE

Perform the operations indicated:

1. $2ax(a^2 + ax + x^2)$.
2. $4mn^2(mn^2 + 3)$.
3. $(16x^2y^2 + 8xy) \div 4xy$.
4. $(a^4 + a^3) \div a^2 + a(a + 1)$.
5. $(36x^3y^2z + 9xy^2z^3) \div 3xyz$.
6. $4a^2x^2 + 9a^2x^2 - 3a^2x^2 + 6a^2x^2$.
7. $32axy - 2axy + y(2ax + 3ax)$.
8. $(m^5 + m^3 + m) \div m + m^2(m^2 + 1) + 1$.
9. $25mnx + 5mnx - 3mnx + 7mnx - 4mnx$.
10. $21p^2q + 42p^2q + p^2(4q + 3q) - (2p^2q - p^2q)$.
11. $(a^2x^2 + ax + 1)a^2x^2 - (a^5x^5 + a^4x^4 + a^3x^3) \div ax$.
12. $6pq(p + q)$. Find the value when $p = 2$, $q = 3$
13. $15pq^2 + 10pq^2 - 5pq^2 + 4pq^2 - (6pq^2 - 3pq^2)$.
14. $4a^2b(a^3 + b^3)$. Find the value when $a = 2$, $b = 1$.
15. $x^2(x^2 + y^2 + z^2)$. Find the value when $x = 1$, $y = 2$, $z = 3$.
16. $(16z^3 + 8z^2 + 12z + 20) \div 4$. Find the value when $z = 1$.
17. $(a^4 + a^3 + a^2 + a) \div a$. Find the value of dividend, divisor, and quotient when $a = 2$.

ORAL EXERCISE

1. $2x \cdot 3y$; $6xy \div 2x$; $6xy \div 3y$.
2. $3a \cdot -2b$; $-6ab \div 3a$; $-6ab \div -2b$.
3. $-4a \cdot 5b$; $-20ab \div -4a$; $-20ab \div 5b$.
4. $-3m \cdot -4n$; $12mn \div -3m$; $12mn \div -4n$.

59. Law of signs.—We see that

Because $+a \cdot +b = +ab$, therefore $+ab \div +a = +b$.
Because $+a \cdot -b = -ab$, therefore $-ab \div +a = -b$.
Because $-a \cdot +b = -ab$, therefore $-ab \div -a = +b$.
Because $-a \cdot -b = +ab$, therefore $+ab \div -a = -b$.

That is,

60. *The quotient of two numbers with like signs is positive; with unlike signs, negative.*

Thus, to divide $32x^2 - 16xy$ by $-16x$, we have

$$-16x)\underline{32x^2 - 16xy}$$
$$-2x + y$$

$32x^2 \div -16x = -2x$, and $-16xy \div -16x = y$.

WRITTEN EXERCISE

1. $45a^2b \div 9b$.
2. $39x^2y^2 \div -13xy$.
3. $275mn \div -25n$.
4. $-625x^2y^2z^2 \div -125xyz$.
5. $-1001x^2y \div 7xy$.
6. $-3535p^2q^2r^2 \div -101p^2r^2$.
7. $(4x^3 - 3x^2) \div -x$.
8. $(9m^2 - 18mn + 27m) \div -9m$.
9. $(-325p^2 + 125q^2) \div -25$.
10. $(17x^3 - 51x^2 + 153x) \div -17x$.

11. Divide $32a^2 - 8ab$ by $-8a$. Then find the value of dividend, divisor, and quotient if $a = -2$, $b = 1$.

12. Divide $49m^3 - 63m^2 - 84m$ by $-7m$. Then find the value of dividend, divisor, and quotient if $m = -1$.

WRITTEN EXERCISE

1. Add $43x^3 + 19x + 27x^2$ and $35 + 172x^2 - 34x^3$.

2. Add $78p^2q - 134pq^2 + 7p^3 - 13q^3$ and $432pq^2 + 48q^3 - 92p^2q$.

3. From $81 + 17x^2 - 43x + 432x^3$ subtract $-42x^4 + 81x^3 - 78 + 37x$.

4. From $923x + 431x^2y - 78xy^3 + 127$ subtract $34 - 92x + 78xy^2 - 227x^2y + 8xy^3$.

5. Multiply $x^6 + 42x^5 - 3x^4 + 81x^3 - 125x^2 + 173x - 144$ by $2x$. Divide the product by $2x$.

6. Multiply $x^5 - 3x^4y + 4x^3y^2 - 81x^2y^3 + 26xy^4 - 17y^5$ by $-3xy$. Divide the product by $-3xy$.

7. Divide $2m^7 + 6m^6n - 18m^5n^2 - 292m^4n^3 + 38m^3n^4 - 16m^2n^5$ by $2m^2$. Multiply the quotient by $2m^2$.

8. Divide $125x^6y^2 + 625x^4y^4 - 325x^3y^5 - 275x^2y^6 + 425x^5y^3$ by $25x^2y^2$. Multiply the quotient by $25x^2y^2$.

Find the value of x *in Exs. 9–21:*

9. $31x = 341$.

10. $127x = 2159$.

11. $176x = 19.36$.

12. $x + 243 = 17$.

13. $243 - x = 2x$.

14. $42x - 4 = 500$.

15. $29x + 3 = 380$.

16. $275 - 2x = 3x$.

17. $34x + 19 = 257$.

18. $47x - 58 = 600$.

19. $\frac{2x}{17} = 48$.

20. $\frac{3x}{11} = 39$.

21. $\frac{17x}{19} = 85$.

22. If 110% of x is 84.70, what is the value of x?

23. If 92% of x is $156.40, what is the value of x?

24. What number increased by 10% of itself equals 143?

25. What number decreased by 7% of itself equals 1953?

26. What number decreased by 9% of itself equals 2912?

FACTORS AND MULTIPLES

ORAL EXERCISE

1. What are the factors of 6? of 10? of 15?

2. What are the factors of $2a$? of $3x$? of ax? of a^7?

3. What are the factors of 9? of 3^2? of a^2? of $35x^3$?

61. Factor. — The word *factor* is used in algebra as it is in arithmetic, the process of separating a quantity into its factors being called *factoring*.

62 How to factor quantities. — We always factor a quantity by thinking of the quantities which must be multiplied together to make it.

The factors of 21 are 3 and 7, because we remember that $3 \cdot 7 = 21$.

The factors of $3a^2$ are 3, a, and a, because we remember that $3\ a\ a = 3a^2$.

The factors of $abx + acx$ are a, x, and $b + c$

Monomials are easily factored at sight.

63. Factoring polynomials. — *To factor a polynomial, examine each term to find the greatest common factor. Then divide to find the other factor.*

For example, to factor $4x^3y + 8x^2y^2 + 20xy^3$, each term contains the factors 2, 2, x, and y.

Hence, dividing by $4xy$, the factors are evidently 2, 2, x, y, and $x^2 + 2xy + 5y^2$.

$$4xy\overline{)4x^3y + 8x^2y^2 + 20xy^3}$$
$$x^2 + 2xy + 5y^2$$

WRITTEN EXERCISE

Factor the following:

1. $9a^3x^2y$; $35x^4y^7z$.
2. $14m^2nx^3y$; $91m^7n^3z$.
3. $mx + my$; $15x^3 + 35x^2y$.
4. $p^2q + pq^2$; $51m^6 + 17m^4x^2$.
5. $3p^3q + 15p^2q^3 + 21p^2q^4$.
6. $x^2y + xy^2 + xz^2 + x^2 + x^3$.

ORAL EXERCISE

1. If a man is worth d dollars and loses rd dollars, how much is he worth? Factor the result.

2. If a man is worth d dollars and gains rd dollars, how much is he worth? Factor the result.

3. If a man has p dollars and gains prt dollars interest, how much has he? Factor the result.

4. If you had n dollars in the bank a year ago, and have gained rn dollars, how much have you in the bank now?

64. Factoring certain formulas. — It is often more convenient to use formulas in factored form. For example, if some goods are marked m and the rate of discount is r, the discount is rm and the selling price is $m - rm$, or $m(1 - r)$. That is,

1. If m = marked price, and
   r = rate of discount,
2. Then rm = the discount, and
   $m - rm$ = the selling price, s.
3. That is, $s = m - rm$
   $= m(1 - r)$.
4. If, now, $m = \$200$, and $r = 10\%$, we have
   $s = \$200(1 - .10)$
   $= \$200 \times .90 = \$180$.

WRITTEN EXERCISE

1. If i, the interest, $= prt$, then the sum of the principal and interest equals $p + prt$. Factor this. Find its value when $p = \$200$, $t = 2$, $r = 5\%$.

2. If s = selling price, c = cost, and r = rate of gain, write a formula for s. Factor it. Find the value of s when $c = \$175$, $r = 20\%$.

ORAL EXERCISE

Name the greatest monomial factor in each polynomial in Exs. 1–10:

1. $ax + aby$.
2. $2\,abc + 4\,bxy$.
3. $3\,pqr^2 + 6\,qrs^2$.
4. $5\,m^2n + 15\,mn^2$.
5. $a^2x + aby + a^3z$.
6. $pqr + qrs + rst$.
7. $a^3m + a^2n + ax + ay$.
8. $m^2x + n^2xy + p^2xz$.
9. $5xy + 10xz + 25x^2 + 50wx$.
10. $7\,xyz + 21\,wxy + 35\,vwx$.

State the products of the quantities in Exs. 11–14:

11. $x(a + b)$.
12. $a(m^2 - n^2)$.
13. $ab(a + b)$.
14. $m^2n^2(m^2 + n^2)$.

State the factors of the following:

15. $ax + bx + cx$.
16. $mx + my + m^2z^2$.
17. $xy + yz + wxy$.
18. $2\,x^2 + 4\,xy + 6\,xz$.
19. $abc + bcd + cde$.
20. $x^3 - 3\,x^2 + 4\,x + xyz$.

WRITTEN EXERCISE

Factor the following:

1. $x^2 - 3\,x$.
2. $m^3 - m^2n$.
3. $a^3 + 2\,a^2b$.
4. $a^3b^2x + a^2by$.
5. $a^2 + a^2b + a^2b^2$.
6. $p^3q^3r - pq^2r^3$.
7. $x^3 + 2\,x^2 + 7\,x$.
8. $p^4 + 3\,p^3q + p^2$.
9. $8\,x^3 + 4\,x^2 + 2\,x$.
10. $x^3 + 15\,x^2 + 16\,x$.
11. $3\,m^4n^3x + 6\,m^2n^2x^2$.
12. $27\,xy^3 + 9\,x^2y^2 + 3\,y$.
13. $a^2b^3c + a^3b^2c + abc^2$.
14. $16\,m^3 + 12\,m^2n + 8\,m^2$.
15. $5\,abcd + 10\,cdef + 15\,c^2d^2$.
16. $17\,a^2 + 51\,abc + 153\,a^3$.
17. $21\,p^2q + 35\,q^2p + 56\,pqr$.
18. $x^4 + 3\,x^3 + 4\,x^2 + 121\,x$.

ORAL EXERCISE

1. Name two multiples of 5; of 7; of a; of pq.

2. Name two common multiples of 5 and 7; of 2 and 8; of a and b.

3. Name the least common multiple of 7 and 9; of 6 and 12; of 8 and 12; of a and b; of ab and bc.

65. Multiple. — As in arithmetic, the product of two quantities is called a *multiple* of either.

For example, ab is a multiple of a and of b.

66. Least common multiple. — The least multiple common to two or more quantities is called their *least common multiple* (l.c.m.).

For example, abc is the least common multiple of ab and bc. In algebra this is often called the *lowest common multiple*.

67. Finding the least common multiple. — The l.c.m. is usually found by factoring.

For example, to find the l.c.m. of $ax + ay$ and $bx + by$, we have: $ax + ay = a(x + y)$, and $bx + by = b(x + y)$; therefore the l.c.m. must contain the factors a, b, $x + y$, and is $ab(x + y)$, or $abx + aby$.

WRITTEN EXERCISE

Find the l.c.m. of the following:

1. abx, bcy.
2. a^2b, ab^2.
3. px^2, qxy.
4. pqr, qrs.
5. mnx^2, $npxy$.
6. ab^2c, a^2bc^2.
7. $a + b$, $ax + bx$.
8. $p^2 + pq$, $q^2 + qp$.
9. $abx + aby$, abc.
10. $27a^3 - 27b^3$, $9ab$.
11. $a(x - y)$, $bx - by$.
12. $x(m + n)$, $y(m + n)$.
13. $41a^2 + 82b^2$, $123ab$.
14. $m^3n + mn^3$, $mp + np$.
15. $am + abn$, $m^2 + bmn$.
16. $x^3 + 3x^2y$, $3m^2y + m^2x$.

WRITTEN EXERCISE

Factor in Exs. 1–22:

1. $39\,a^2bc + 65\,abc^2$.	**2.** $m^3n^2p + mn^2p^3$.
3. $17\,u^2v - 119\,uv^2$.	**4.** $pqr - qrs + rst$.
5. $x^3 + 3\,x^2y + 5\,xy^2$.	**6.** $252\,x^2y^3 + 84\,xy$.
7. $a^4 + 3\,a^3b - 4\,a^2b^2$.	**8.** $86\,x^3y + 129\,x^2y^2$.
9. $abcd + bcde + defg$.	**10.** $x^3 - 3\,x^2y + 3\,xy^2$.
11. $27\,amn + 108\,mnp^2q$.	**12.** $57\,m^3n^2 + 95\,m^2n^3$.
13. $m^2np + mn^2p - mnp^2$.	**14.** $m^4 + m^2n + 4\,m^3n^2$.
15. $p^2 - 3\,p^3 - 4\,p^4 + 5\,p$.	**16.** $6\,x^2 + 8\,xy + 10\,x^3$.
17. $-p^2q^3r - pq^2r^3 - pqr$.	**18.** $82\,p^3q^2r - 123\,p^2q^2r^3$.
19. $42\,a^2b - 63\,ab^2 + 168\,abc$.	**20.** $32\,x^3 + 72\,x^2y^2 - 128\,x$.
21. $37\,p^2q + 185\,pqr - 74\,pq$.	**22.** $-8\,x^3y^2z - 6\,x^2y^3z - 4\,xyz$

Find any multiple of the quantities in Exs. 23–31:

23. p^2qr.	**24** $3\,q^2rz$.	**25.** $7\,x^2y^2z$.
26. $a + b$.	**27.** $x - y$.	**28.** $m^2 - n^2$.
29. $a^2 + b^2 + c^2$.	**30.** $a^2 + 2\,ab + b^2$.	**31.** $a - b + c$.

Find any common multiple in Exs. 32–37:

32. ab, $3\,bc$.	**33.** $2\,pq$, $3\,qr$.	**34.** m^3, n^3, x^2.
35. a^2, $a^2 - b^2$.	**36.** m^2, $m + n$.	**37.** $2\,a^2b$, $3\,ab^2$.

Find the l.c.m. in Exs. 38–49:

38. $2\,a^2b$, $4\,ab^2$.	**39.** $a^2 - b^2$, $15\,ab$.
40. $a + b$, $25\,ab$.	**41.** $p^2 + 2\,p$, $27\,p^3$.
42. ab, bc, cd, de.	**43.** $15\,p^2q^2r$, $20\,pqr$.
44. $32\,abc$, $2\,a + b$.	**45.** $2\,a$, $3\,b$, $4\,c$, $5\,d$.
46 abc, bcd, $a + b + c$.	**47.** $15\,p^2q^2r^2$, $p + q + r$.
48. $a^2 + 2\,ab + b^2$, $3\,a^2$.	**49** $2\,a^2 - 6\,ab + 2\,b^2$, $2\,a$.

FRACTIONS

ORAL EXERCISE

1. What are the terms of the fraction $\frac{2}{3}$? Which is the numerator? Which is the denominator? Answer the same questions for the fraction $\frac{a}{b}$.

2. In the fraction $\frac{3}{4}$, into how many equal parts has the unit been divided, and how many have been taken? Answer the same questions for the fraction $\frac{a}{b}$.

3. If we think of $\frac{15}{10}$ as an expression of division, which number is the dividend? Which is the divisor? Answer the same questions for the fraction $\frac{x}{y}$.

68. A fraction. — One or more of the equal parts of any unit is called a *fraction.*

A fraction may also be considered as an expression of division.

Thus, $\frac{2}{3}$ means 2 of the 3 equal parts of 1, or it means that 2 has been divided into 3 equal parts. $\frac{a}{b}$ means a of the b equal parts of 1, or it means that a has been divided into b equal parts.

69. Terms of a fraction. — The terms *numerator* and *denominator* are used as in arithmetic.

70. Integer. — An algebraic quantity in which no fraction is expressed is said to be an *integer,* or to be *integral.*

For example, $ax + b$ is an integer, and is integral.

71. Sign of the fraction. — The sign written before the fraction is called the *sign of the fraction.*

For example, $-\frac{2}{3}$ is a negative fraction, and $\frac{3a}{4b}$ is a positive fraction.

ORAL EXERCISE

1. In reducing $\frac{15}{20}$ to lowest terms what factor is canceled? What is the result? Answer these questions for $\frac{ax}{ay}$.

2. Reduce each of these fractions to lowest terms:

$$\frac{m^2x}{m^2y} \qquad \frac{abx^2}{aby^2} \qquad \frac{4\,x^2y}{6\,xy} \qquad \frac{14\,p^2qr}{21\,p^2qs}$$

72. Reducing fractions to lowest terms. — Fractions are treated in algebra just as in arithmetic. *Fractions are reduced to lowest terms by canceling all factors common to numerator and denominator.*

Thus, the fraction $\frac{x^2 + xy}{2\,x}$ is reduced to $\frac{x+y}{2}$ by canceling the only common factor, x.

WRITTEN EXERCISE

Reduce the following to lowest terms:

1. $\frac{a^2b}{b^2a}$.
2. $\frac{a^2b^2x^2}{b^2x^2y^2}$.
3. $\frac{a^3xy}{a^2x^2y}$.
4. $\frac{ab^2c^3}{a^3b^2c}$.
5. $\frac{21\,a^2xy}{35\,ax^2y^2}$.
6. $\frac{16\,a^3x^2y}{36\,a^7yz^3}$.
7. $\frac{ax^2 + ay^2}{3\,a^2}$.
8. $\frac{x^2y + y^2x}{axy}$.
9. $\frac{p^2qr + q^2rs}{pqrs}$.
10. $\frac{2\,a^2 + 3\,ab}{4\,a}$.
11. $\frac{51\,m^2n + 85\,mn^2}{17\,mn}$.
12. $\frac{3\,a^2 + 3\,ab}{21\,ax}$.
13. $\frac{a^3 + 3\,a^2b + 3\,ab^2}{ab}$.
14. $\frac{p^4q + p^3q^2 + p^2q^3 + pq^4}{pqr}$.
15. $\frac{m^2n + mn + mn^2}{mn}$.
16. $\frac{81\,x^4y + 108\,x^2y^2 + 207\,xy^4}{27\,xy}$.

WRITTEN EXERCISE

1. If the perimeter of a square is 16, what is each side? What is the area? If the perimeter is p, what is each side? What is then the area?

2. If the perimeter of a rectangle 16 in. long is 50 in., what is the area? Suppose the perimeter is p and the length is l?

3. If a piece of silk of a yards is worth b dollars, how much will an employee of the store have to pay for a yard, allowing him a discount of c cents a yard? What is the result if $a = 40$, $b = 60$, $c = 25$?

4. A car wheel is 6 ft. in circumference. How many revolutions does it make in going 60 ft.? in going d ft.? If it is c ft. in circumference, the number of revolutions, r, in going d ft., is how many? That is, what does r equal?

5. If a barrel weighing w lb. is rolled up an incline s ft. long, to a point d ft. high, a power of $\frac{wd}{s}$ lb. is exerted; that is, $p = \frac{wd}{s}$. How much power must be used to roll a 200-lb. barrel up a 10-ft. incline to a height of 4 ft.?

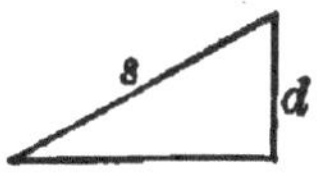

6. How much power must be used to roll a 100-lb. barrel up an 8-ft. incline to a height of 4 ft.? also a 150-lb. barrel up a 12-ft. incline to a height of 2 ft.? also a 300-lb. barrel up a 15-ft. incline to a height of 5 ft.?

7. Two cogwheels, one having 9 cogs and the other 27, are fitted together. How many times will the smaller wheel turn for each turn of the larger? How many times, if the larger has 10 cogs and the smaller 5? if the larger has a and the smaller b?

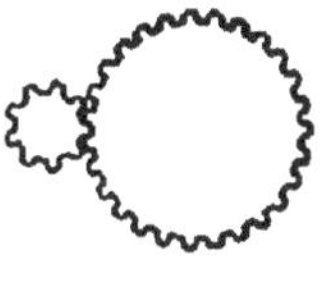

ORAL EXERCISE

1. How do we reduce $\frac{2}{3}$ to sixths? $\frac{a}{b}$ to $b c$ths?

2. Reduce to fractions with denominator abx: $\frac{c}{x}, \frac{a}{b}, \frac{m}{ab}$.

3. Reduce to fractions having the least common denominator: $\frac{2}{3}$ and $\frac{4}{5}$; $\frac{2}{3}$ and $\frac{5}{6}$; $\frac{a}{b}$ and $\frac{c}{d}$.

73. Reduction of fractions. — As in arithmetic,

Both terms of a fraction may be multiplied, or both divided, by the same quantity without changing the value of the fraction.

For example, $\frac{2}{3} = \frac{10}{15}, \quad \frac{25}{30} = \frac{5}{6}, \quad \frac{a}{b} = \frac{xa}{xb}, \quad \frac{ax}{xy} = \frac{a}{y}$.

74. Least common denominator — The least denominator common to several fractions is called their *least common denominator* (l.c.d.).

This is also, in algebra, called the *lowest common denominator.*

For example, the l.c.d. of $\frac{a}{b}, \frac{c}{x}$, and $\frac{m^2}{c^2}$ is bxc^2.

75. Finding the l.c.d. — As in arithmetic, the l.c.d. is evidently the l.c.m. of the denominators.

For example, to reduce $\frac{a}{x+y}, \frac{a}{2b}$, and $\frac{m}{4b^2}$ to fractions having the l.c.d.

The l.c.d. must evidently contain the factors $x + y$, 2, b, 4 (which contains 2), and b^2 (which contains b). It is therefore $4b^2(x+y)$. Multiplying both terms of $\frac{a}{x+y}$ by $4b^2$, of $\frac{a}{2b}$ by $2b(x+y)$, and of $\frac{m}{4b^2}$ by $x + y$, we have the results.

$$\frac{a}{x+y} = \frac{4ab^2}{4b^2(x+y)}$$
$$\frac{a}{2b} = \frac{2ab(x+y)}{4b^2(x+y)}$$
$$\frac{m}{4b^2} = \frac{m(x+y)}{4b^2(x+y)}$$

WRITTEN EXERCISE

Reduce to fractions having the denominator indicated in parentheses in Exs. 1–10:

1. $\frac{a}{2b}$, $(4b^3)$.

2. $\frac{p}{3q^2r}$, $(12q^2r^2)$.

3. $\frac{m}{n^2p^2}$, $(m^2n^3p^4)$.

4. $\frac{x}{p^2qr^3}$, $(2p^2q^2r^5)$.

5. $\frac{a}{b+c}$, (b^2+bc).

6. $\frac{p}{q-r}$, $(2q^2-2qr)$.

7. $\frac{a+b}{a-b}$, $(axy-bxy)$.

8. $\frac{x^2-1}{x^2+1}$, $(x^2y^2z^2+y^2z^2)$.

9. $\frac{m}{m^2-n^2}$, $(ab^2m^2-ab^2n^2)$.

10. $\frac{x}{w^2-2wx+x^2}$, $(3w^2+3x^2-6wx)$.

Reduce to fractions having the l.c.d.:

11. $\frac{a}{b}, \frac{b}{c}, \frac{c}{a}$.

12. $\frac{a}{2b}, \frac{m}{2n}, \frac{x}{bn}$.

13. $\frac{1}{a}, \frac{1}{a^2}, \frac{1}{a^3}$.

14. $\frac{p}{q^2}, \frac{q}{r^2}, \frac{1}{p^2q^2r^2}$.

15. $\frac{p}{q+r}, \frac{1}{q}, \frac{1}{r}$.

16. $\frac{x}{y^2+z^2}, \frac{x}{y}, \frac{x}{z}$.

17. $\frac{a}{p-q+r}, \frac{n}{m}$.

18. $\frac{1}{pqr}, \frac{2}{qrs}, \frac{3}{rst}$.

19. $\frac{a^2+b^2}{a^2-b^2}, \frac{a}{b}, \frac{b}{a}$.

20. $\frac{a^2-b^2}{a^2+b^2}, \frac{a}{b}, \frac{a^2}{b^2}$.

21. $\frac{a+b+c}{a-b+c}, \frac{a}{b}, \frac{b}{c}$.

22. $\frac{ab}{3pqr}, \frac{bc}{6qrs}, \frac{cd}{9pqs}$.

23. $\frac{1}{a-b-c}, \frac{b}{a}, \frac{c}{b}$.

24. $\frac{x}{2a+2b}, \frac{y}{2ab}, \frac{z}{4c}$.

ORAL EXERCISE

1. How many fifths in 1? in $1\frac{3}{5}$? in 2? in $2\frac{4}{5}$?
2. How many $b$ths in 1? in 3? in m? in $m + \frac{a}{b}$?
3. Reduce a to $x$ths; $a + \frac{x}{y}$ to $y$ths; $x^2 + \frac{2}{y^2}$ to y^2ths.

76. **Mixed quantities.** — As in arithmetic we speak of $2\frac{3}{4}$ as a *mixed number*, so in algebra $a + \frac{b}{c}$ is a *mixed quantity*.

77. **Reduction to fractional forms.** — Mixed quantities in algebra are reduced in the same way as in arithmetic.

Arithmetic:	*Algebra:*
Reduce $3\frac{4}{5}$ to fifths.	Reduce $x + \frac{n}{d}$ to $d$ths.
1. Since $1 = \frac{5}{5}$, therefore $3 = \frac{15}{5}$.	1. Since $1 = \frac{d}{d}$, therefore $x = \frac{xd}{d}$.
2. $\frac{15}{5} + \frac{4}{5} = \frac{19}{5}$.	2. $\frac{xd}{d} + \frac{n}{d} = \frac{xd + n}{d}$.

WRITTEN EXERCISE

Reduce to fractional forms:

1. $x + \frac{p}{q}$ to $q$ths.
2. $3m^2 + \frac{n^2}{k}$ to $k$ths.
3. $2a^2b + c + \frac{a}{b}$ to $b$ths.
4. $a^2 - 3ab + b^2$ to $b$ths.
5. $m^2 + 3mn + 2n^2$ to $n$ths.
6. $x^2 + 2xy + y^2 + \frac{w}{y}$ to $y$ths.
7. $3p^3 + \frac{p^2}{2q}$.
8. $4m^2n + \frac{mn}{3x}$.
9. $4a + 2b + \frac{c}{d^2}$.
10. $16x^2 + 15xy + \frac{3x}{y}$.
11. $32x^2 - 15y^2 + \frac{x}{y}$.
12. $37a^2 - 15b^3 + \frac{16b}{5a}$.

ORAL EXERCISE

1. Reduce to whole numbers: $\frac{8}{2}$; $\frac{16}{8}$; $\frac{2a}{a}$; $\frac{5x^2}{5}$.

2. Reduce to whole numbers: $\frac{a(b+c)}{a}$; $\frac{3m^2(m+n)}{3m^2}$.

3. Reduce to mixed numbers: $\frac{5}{2}$; $\frac{7}{4}$; $\frac{a+1}{a}$; $\frac{x+3}{x}$.

78. Remainders in division. — In algebra, as in arithmetic, a remainder in division leads to a fraction in the quotient.

Arithmetic:	*Algebra:*
Divide 124 by 5.	Divide x^2+3x+1 by x.
$5\overline{)124}$	$x\overline{)x^2+3x+1}$
24, 4 remainder,	$x+3$, 1 remainder,
or $24\frac{4}{5}$.	or $x+3+\frac{1}{x}$.

79. Reduction of improper fractions. — *To reduce an improper fraction to a whole or a mixed expression, divide the numerator by the denominator.*

For example,

$$\frac{x^3+2x^2+3x+1}{x} = x^2+2x+3+\frac{1}{x}.$$

$$x\overline{)x^3+2x^2+3x+1}$$
$$x^2+2x+3+\frac{1}{x}$$

WRITTEN EXERCISE

Reduce to whole or mixed quantities:

1. $\frac{2x^3+5x^2+1}{x}$.

2. $\frac{p^2+q+r}{p}$.

3. $\frac{16m^2+32m+7}{16m}$.

4. $\frac{4a^4+6a^2+3}{2a^2}$.

5. $\frac{8x^3+4x^2+2x+6}{2x}$.

6. $\frac{12a^3+6a^2b+17b^2}{6a^2}$.

WRITTEN EXERCISE

Reduce to fractions having the denominators indicated:

1. $\frac{2m}{7}$, (21). 2. $\frac{4xy}{5}$, (30). 3. $\frac{2p^2q}{3}$, (15).

4. $\frac{14m^2n}{9}$, (81). 5. $\frac{a^2bc}{3x^2yz}$, $(21x^2y^2z^2)$. 6. $\frac{7q^2rx^2y}{9z}$, $(18xyz)$.

7. $\frac{2a+4b}{7}$, (49). 8. $\frac{p^2-4q}{5}$, $(25x)$. 9. $\frac{2xy+y^2}{xy}$, (xyz).

Reduce to fractions in their lowest terms in Exs. 10–18:

10. $\frac{4pq}{16p^2q^2}$. 11. $\frac{21a^2b}{49b^2c}$. 12. $\frac{17m^2n^3}{51m^3n^2}$.

13. $\frac{125abc^2}{225a^2bc}$. 14. $\frac{2p^2+3p}{3p^2+2p}$. 15. $\frac{2x}{x^2+x}$.

16. $\frac{x^3+2x}{x^3+3x}$. 17. $\frac{35x+35y}{49}$. 18. $\frac{3m^3}{9m^2+27m}$.

Reduce to whole or mixed quantities in Exs. 19–27:

19. $\frac{27x^2y^2}{9xy}$. 20. $\frac{225a^3b^2}{25ab}$. 21. $\frac{625p^5q^4}{125p^4q^4}$.

22. $\frac{45ab+7}{9a}$. 23. $\frac{81x^2y+19}{9xy}$. 24. $\frac{3x^3+5}{3x}$.

25. $\frac{24x^2-2y}{6x}$. 26. $\frac{32x^2y^2+15}{8xy}$. 27. $\frac{27x^2+81x+25}{9x}$.

Write the following in fractional form:

28. $a+\frac{2b}{4c}$. 29. $p^2+\frac{4q}{5p^2}$.

30. $a+b+\frac{c}{d}$. 31. $x^2+2xy+\frac{4}{y}$.

32. $21x^4+17x^3+\frac{1}{x^2}$. 33. $82x^2y+21xy^2+\frac{x}{y}$.

ORAL EXERCISE

1. If half of your weight is 40 lb., how much do you weigh? If $\frac{1}{2}$ of x is 40, how much is x?

2. Find the value of x in each of the following statements:

$$\frac{x}{2}=40 \qquad \frac{x}{2}=15 \qquad \frac{x}{2}=50 \qquad \frac{x}{3}=10 \qquad \frac{x}{5}=6$$

3. If $\frac{x}{2}+1=11$, what does $\frac{x}{2}$ equal? What does x equal? If $\frac{x}{2}-1=4$, what does $\frac{x}{2}$ equal? What does x equal?

4. If you have equal weights in the two pans of the scales, will they balance if you add 2 oz. to each? subtract 2 oz. from each? multiply each by 2? divide each by 2? State four operations which you may perform on the two members of an equation without destroying the equality.

80. Multiplying equals by equals to avoid fractions. — In the equation $\frac{x}{3}=7$, if we multiply these equals by 3 we have $x=21$. In the equation $\frac{2x}{5}+6=8$ we have $\frac{2x}{5}=2$, or $\frac{x}{5}=1$; multiplying by 5, $x=5$.

81. Clearing an equation of fractions. — Multiplying both members by such a number as to make fractional terms integral is called *clearing the equation of fractions.*

WRITTEN EXERCISE

Find the value of x:

1. $\frac{x}{3}+2=9.$ 2. $\frac{x}{5}+2=30.$ 3. $\frac{2x}{7}+5=15.$

4. $\frac{3x}{7}+4=13.$ 5. $\frac{5x}{6}-7=13.$ 6. $\frac{9x}{11}-8=10.$

7. $42+\frac{x}{7}=86.$ 8. $37+\frac{2x}{5}=63.$ 9 $17+\frac{5x}{6}=67.$

I am thinking of a number whose half added to 71 amounts to 97. What is the number?

1. If $x =$ the number,

2. Then $\frac{x}{2} =$ half of the number.

3. Then $\frac{x}{2} + 71 = 97$, by the statement.

4. Then $\frac{x}{2} = 26$, by taking 71 from these two equals, and

5. $x = 52$, by multiplying these equals by 2.

Check. If I add 71 to 26 (half of 52), the sum is 97.

82. Checking the result. — *Always check by placing the result in the original statement. You may have made a mistake in getting your equation as well as in solving.*

WRITTEN EXERCISE

1. What is that number $\frac{2}{3}$ of which, plus 5, is 45?

2. What is that number $\frac{4}{5}$ of which, less 6, is 10?

3. If I add 10 to 5% of a certain number, the result is 30. What is the number?

4. I am thinking of a number such that its seventh less 6 is 27. What is the number?

5. If I subtract 6 from 7% of a certain number, the result is 15. What is the number?

6. If to a certain number I add half of the same number, and then add 7, the result is 10. What is the number?

7. If from a certain number I take $\frac{1}{3}$ of the same number, and then add 10, the result is 30. What is the number?

8. Make up a problem somewhat like the above, and then solve it.

Teachers will find that the pupils will derive much benefit from making original problems as here suggested.

ADDITION OF FRACTIONS

ORAL EXERCISE

1. Add $\frac{2}{7}$ and $\frac{3}{7}$; $\frac{a}{7}$ and $\frac{b}{7}$; $\frac{a}{c}$ and $\frac{b}{c}$.

2. Add $\frac{2}{x}$ and $\frac{7}{x}$; $\frac{2a}{x}$ and $\frac{3a}{x}$; $\frac{a+b}{x}$ and $\frac{a}{x}$.

83. Addition of fractions. — Fractions are added in algebra in the same way as in arithmetic, by first reducing to the least common denominator.

Arithmetic:

Add $\frac{3}{4}$ and $\frac{1}{6}$.

1. The least common denominator is evidently 12.

2. $$\frac{3}{4} = \frac{9}{12},$$
and $$\frac{1}{6} = \frac{2}{12}.$$

3. The sum $= \frac{11}{12}$.

Algebra:

Add $\frac{a}{bc}$ and $\frac{c}{bd}$.

1. The least common denominator is evidently bcd.

2. $$\frac{a}{bc} = \frac{ad}{bcd},$$
and $$\frac{c}{bd} = \frac{c^2}{bcd}.$$

3. The sum $= \frac{ad + c^2}{bcd}$.

WRITTEN EXERCISE

1. $\frac{a}{b} + \frac{c}{d}$.

2. $\frac{m}{n} + \frac{n}{m}$.

3. $\frac{abc}{xyz} + \frac{abc}{wxy}$.

4. $\frac{a^2bc}{2p^2q} + \frac{b^2cd}{3pq^2}$.

5. $\frac{3a^2b}{4m^2a} + \frac{4ab^2}{3mn^2}$.

6. $\frac{a^2-b^2}{4ac} + \frac{4b^2}{a^2}$.

7. $\frac{a+b}{c} + \frac{a-2b}{2c}$.

8. $\frac{5x^8y^2}{6mn} + \frac{17x^3y^2}{36mn}$.

SUBTRACTION OF FRACTIONS

ORAL EXERCISE

1. $\frac{5}{7}-\frac{2}{7}$; $\frac{5}{a}-\frac{2}{a}$; $\frac{3a}{b}-\frac{a}{b}$; $\frac{a}{x}-\frac{b}{x}$; $\frac{m}{p^2}-\frac{n}{p^2}$.

2. $\frac{1}{2}-\frac{1}{4}$; $\frac{1}{2}-\frac{3}{8}$; $\frac{3a}{2b}-\frac{a}{b}$; $\frac{15x}{2y}-\frac{3x}{y}$; $\frac{a}{m}-\frac{b}{3m}$.

84. Subtraction of fractions. — Fractions are subtracted in algebra in the same way as in arithmetic.

Arithmetic:	*Algebra:*
From $\frac{3}{4}$ take $\frac{1}{6}$.	From $\frac{a+b}{2c}$ take $\frac{a-b}{bc}$.
1. The least common denominator is evidently 12.	1. The least common denominator is evidently $2bc$.
2. $\frac{3}{4}=\frac{9}{12}$,	2. $\frac{a+b}{2c}=\frac{ab+b^2}{2bc}$,
and $\frac{1}{6}=\frac{2}{12}$.	and $\frac{a-b}{bc}=\frac{2a-2b}{2bc}$.
3. The difference $=\frac{7}{12}$.	3. The diff. $=\frac{ab+b^2-2a+2b}{2bc}$.

85. It should be noticed that the fraction bar has the same effect as parentheses, the $2a-2b$ in the above example being subtracted as explained on page 37.

WRITTEN EXERCISE

1. $\frac{a}{b}-\frac{c}{d}$

2. $\frac{a}{bc}-\frac{c}{ab}$.

3. $\frac{2a}{3x}-\frac{4b}{6y}$.

4. $\frac{5ab}{6c^2d}-\frac{3ab}{4cd^2}$.

5. $\frac{a}{b}-\frac{a-b}{b}$.

6. $\frac{x}{3mn^2}-\frac{x-y}{2m^2n}$.

WRITTEN EXERCISE

Add and subtract as indicated:

1. $\frac{x}{yz} + \frac{y}{xz}$.

2. $\frac{m}{pq} - \frac{p}{mq}$.

3. $\frac{a^2b}{c^2d} + \frac{ab^2}{cd^2}$.

4. $\frac{m^2n}{p^2q} - \frac{mn^2}{pq^2}$.

5. $\frac{a+b}{3\,ab^2} + \frac{a-b}{3\,a^2b}$.

6. $\frac{ax}{by} + \frac{bx}{ay} - \frac{x}{aby}$.

7. $\frac{a}{x^2} + \frac{b}{y^2} - \frac{a-b}{x^2y^2}$.

8. $\frac{m+n}{7\,m} - \frac{m-n}{35\,n}$.

9. $\frac{a^2+1}{a^2bc} + \frac{a^2-1}{ab^2c}$.

10. $\frac{p^2+q^2}{pq} - \frac{q^2+r^2}{qr}$.

11. $\frac{2\,p}{3\,q} + \frac{4\,p}{5\,q} - \frac{7\,p}{15\,q}$.

12. $\frac{a}{b^2} + \frac{b}{a^2} - \frac{a^3+b^3}{a^2b^2}$.

13. $\frac{x^2-ay^2}{abc} - \frac{x^2-dy^2}{bcd}$.

14. $\frac{x^2+y^2}{2\,x} - \frac{x^2-y^2}{4\,y}$.

15. $\frac{a+b+c}{abc} - \frac{b+c+d}{bcd}$.

16. $\frac{a}{b} - \frac{b}{c} + \frac{c}{d} - \frac{d}{e} + \frac{e}{f}$.

17. $\frac{x+y}{x} + \frac{y+z}{y} + \frac{z+x}{z}$.

18. $\frac{a+b}{ab} + \frac{b+c}{bc} + \frac{c+a}{ca}$.

19. $\frac{a^2+2\,ab+b^2}{ab} - \frac{a^2-2\,ab+b^2}{ab}$.

20. $\frac{a+1}{b} - \frac{a-1}{b^2} - \frac{1-a}{b^3}$.

21. $\frac{x^3-1}{y} + \frac{y^3-1}{z} + \frac{z^3-1}{x}$.

22. $\frac{1-a+a^2}{b} - \frac{1-b+b^2}{a}$.

23. $\frac{-p-q-r}{pq} - \frac{p+q+r}{qr}$.

24. $\frac{a+b}{c} + \frac{b+c}{a} - \frac{b(a+c)}{ac}$.

25. $\frac{abc}{uvw} + \frac{bcd}{vwx} + \frac{cde}{wxy} + \frac{def}{xyz}$.

26. $\frac{a^2+b^2}{a^2b^2} + \frac{b^2+c^2}{b^2c^2} + \frac{c^2+a^2}{c^2a^2}$.

27. $\frac{a+b+c+d}{x^2y^2} + \frac{b+c+d+e}{y^2z^2} + \frac{c+d+e+f}{z^2w^2}$.

MULTIPLICATION OF FRACTIONS

ORAL EXERCISE

1. How much is twice $\frac{1}{3}$? a times $\frac{1}{b}$? m times $\frac{x}{y}$?

2. How much is 5 times $\frac{2}{15}$? (Cancel.) m times $\frac{x}{my}$?

3. How much is 10 times $\frac{2}{15}$? (Cancel.) mn times $\frac{x}{my}$?

4. Tell how to multiply a fraction by an integer.

86. Multiplication of unit fractions. — It appears from the above exercise that we multiply fractions in algebra in the same way as in arithmetic.

For example, consider unit functions:

Arithmetic:

Take $\frac{1}{3}$ of $\frac{1}{5}$.

We think of 1 as divided into 5 equal parts, and of each part as divided into 3 equal parts, so that 1 is divided into 5×3, or 15, equal parts.

Therefore $\frac{1}{3}$ of $\frac{1}{5} = \frac{1}{15}$.

Algebra:

Take $\frac{1}{b}$ of $\frac{1}{d}$.

We think of 1 as divided into d equal parts, and of each part as divided into b equal parts, so that 1 is divided into $b \cdot d$, or bd, equal parts.

Therefore $\frac{1}{b}$ of $\frac{1}{d} = \frac{1}{bd}$.

WRITTEN EXERCISE

1. $a \cdot \frac{1}{x}$.

2. $-b^2 \cdot \frac{-b}{xy}$.

3. $a^2b \cdot \frac{ab^2}{c^2d^2}$.

4. $\frac{1}{m} \cdot \frac{x+y}{7xy}$.

5. $\frac{1}{n} \cdot \frac{a-b}{16xy}$.

6. $-a \cdot \frac{a-b}{cd}$.

ORAL EXERCISE

1. How much is $\frac{1}{b}$ of $\frac{1}{d}$? $\frac{1}{b}$ of $\frac{c}{d}$? $\frac{a}{b}$ of $\frac{c}{d}$?

2. How do you multiply one fraction by another?

87. Multiplication of fractions. — Compare arithmetic and algebra again, as on page 68.

Arithmetic:

Take $\frac{2}{3}$ of $\frac{4}{5}$.

Because 4 has been divided into 5 equal parts, and each of these into 3, therefore 4 has been divided into 15 equal parts.

Therefore $\frac{1}{3}$ of $\frac{4}{5} = \frac{4}{15}$.

Therefore $\frac{2}{3}$ of $\frac{4}{5} = 2 \cdot \frac{4}{15} = \frac{8}{15}$.

Algebra:

Take $\frac{a}{b}$ of $\frac{c}{d}$.

Because c has been divided into d equal parts, and each of these into b, therefore c has been divided into bd equal parts.

Therefore $\frac{1}{b}$ of $\frac{c}{d} = \frac{c}{bd}$.

Therefore $\frac{a}{b}$ of $\frac{c}{d} = a \cdot \frac{c}{bd} = \frac{ac}{bd}$.

88. Therefore in algebra, as in arithmetic,

To multiply one fraction by another, multiply the numerators for a new numerator and the denominators for a new denominator.

89. Cancellation should be employed whenever possible, as in the case of $\frac{a}{b} \cdot \frac{x+y}{a} = \frac{x+y}{b}$.

WRITTEN EXERCISE

1. $\frac{a}{b^2} \cdot \frac{c^2}{d}$.
2. $\frac{-a^2}{b} \cdot \frac{-c}{d^2}$.
3. $\frac{a}{m} \cdot \frac{b+c}{n}$.
4. $\frac{a^2bc}{m^2nx} \cdot \frac{ab^2c^2}{mn^2x^2}$.
5. $\frac{-2ab^2}{3xyz} \cdot \frac{-3xy}{2ab}$.
6. $\frac{a^8b^2c}{xy^2z^8} \cdot \frac{x^8y^2z}{ab^2c^8}$.

ORAL EXERCISE

1. At m miles in h hours, how far will a train go in 1 hr.? in t hr.? in $2h$ hr.? in h^2 hr.?

2. At m miles in h hours, how far can you walk in $\frac{a}{b}$ of an hour? in $3h$ hr.? in $25h$ hr.?

3. If y yards cost d dollars, what will 1 yd. cost? $\frac{m}{n}$ yd.? m yd.?

4. A man earns d dollars a week for working 6 days, h hours a day. How much is this per day? per hour?

5. In a circle of circumference c and diameter d, $\frac{c}{d} = \frac{22}{7}$. What is the value of $\frac{\frac{1}{2}c}{d}$? of $\frac{\frac{1}{11}c}{d}$? of $\frac{7c}{d}$?

6. If a acres of land are worth d dollars, how much is 1 acre worth? $\frac{2}{3}$ of an acre? $3a$ acres? $25a^2$ acres?

WRITTEN EXERCISE

1. A man had $\frac{a}{b}$ of an acre in a village lot, and bought $\frac{c}{d}$ of an acre adjoining. How much was his land then worth at d dollars an acre?

2. What is the value of the land in Ex. 1, if $a = 1$, $b = 2$, $c = 300$, $d = 400$? (Simply substitute in the result.)

3. One automobile can go m miles in h hours, and another can go $\frac{a}{b}$ as fast. What is the rate of the second one per hour?

4. What is the value of the result in Ex. 3, if $m = 63$, $h = 3$, $a = 2$, $b = 3$?

5. If n pupils are divided equally in c classes, and if half are boys, how many boys in each class?

WRITTEN EXERCISE

1. $\frac{a}{b}\cdot\frac{b^2}{c^2}\ \frac{c^8}{d^8}\cdot\frac{d^4}{a^4}.$

2. $\frac{a^2b}{c^2d}\cdot\frac{cd^2}{ef^2}\ \frac{e^2f}{ab^2}.$

3. $\frac{m^2n}{p^2q^2r^2}\cdot\frac{p^2q}{m^2n^2s^2}.$

4. $\frac{w^4x^8}{y^4z^8}\ \frac{y^2z}{w^2x}\cdot\frac{w}{x}\cdot\frac{y}{z}.$

5. $\frac{a+b}{x^2y}\cdot\frac{x}{y^2z}\ \frac{xy^8z}{2}.$

6. $\frac{a-b}{mn}\cdot\frac{ab}{m^2n^2}\cdot\frac{m^4n^4}{3}.$

7. $\frac{3\,p^4q^8r^2}{4\,ab^2c^3}\ \frac{6\,a^4b^8c^2}{5\,pq^2r^8}.$

8. $\frac{a^8b}{-c^8d}\ \frac{b^8c}{-d^8e}\ \frac{c^8d}{-e^8a}.$

9. $\frac{-x}{y}\ \frac{-y}{z}\ \frac{-z}{x}\cdot\frac{p}{q}.$

10. $\frac{17\,a^5b^7c^8}{37\,m^2n}\cdot\frac{259\,m^6n^9}{119\,a^9b^{11}c^{16}}.$

11. $\frac{-2}{a}\ \frac{-3}{b}\ \frac{-4}{c}\cdot\frac{-abc}{48}.$

12. $\frac{a}{b}\cdot\frac{b}{c}\cdot\frac{-c}{d}\cdot\frac{-d}{e}\cdot\frac{e}{-f}\ \frac{b}{-a}.$

13. $\frac{a^8b^2}{c^8d^2}\cdot\frac{-b^8c^2}{d^8e^2}\cdot\frac{c^8d^2}{-e^8a^2}\ \frac{-2}{b^8c^2}.$

14. $\frac{x}{-u}\ \frac{y}{-v}\ \frac{z}{-w}\ \frac{-u\ -v\ -u}{xyz}.$

15. $\frac{-1}{mn}\ \frac{-1}{mn}.$

16. $a^2b\ \frac{a^2-b}{m^2n^2}.$

17. $-\frac{1}{x}\cdot\frac{-1}{y}.$

18. $\frac{1}{ab^2cd^2}\cdot\frac{1}{a^2bc^2d}.$

19. $\frac{-1}{2\,pq^2r^3}\ \frac{-1}{3\,p^8q^2r}.$

20. $\frac{1}{a^4b^3x^2y}\ \frac{1}{ab^2x^8y^4}.$

21. $\frac{a^2-b^2}{a^2b^2}\ \frac{ab}{-cd}\cdot\frac{-c^4d^2}{ab^8}\ \frac{a^2b^4}{37}.$

22. $\frac{p+q+r+s}{pq}\ \frac{rs}{p^2q^2}\cdot\frac{p^8q^8}{r^2s^2}\ \frac{-rs}{8}.$

23. $\frac{x^2-2\,xy+y^2}{a^4b^8c^2}\cdot\frac{x^4y^8z^2}{-ab^2c^8}\ \frac{-a^5b^5c^5}{x^8y^2z}.$

24. $\frac{a^2+b^2+c^2}{abc}\ \frac{-pqr}{xyz}\cdot\frac{-a^2x^2b^2y^2c^2z^2}{p^2q^2r^2}.$

DIVISION OF FRACTIONS

ORAL EXERCISE

1. Because $2 \cdot 3 = 6$, we know that $6 \div 2 =$ how many? Because $a \cdot b = ab$, we know that $ab \div a =$ how many?

2. Because $\frac{a}{b} \cdot \frac{c}{d} = \frac{ac}{bd}$, we know that $\frac{ac}{bd} \div \frac{a}{b} =$ how many?

90. Division of fractions. — Because $\frac{ac}{bd} \div \frac{a}{b} = \frac{c}{d}$, and also $\frac{ac}{bd} \cdot \frac{b}{a} = \frac{c}{d}$, we see that

The result of dividing by a fraction is the same as that of multiplying by the fraction inverted.

Thus, to divide $\frac{m}{n}$ by $\frac{x}{y}$, we may multiply $\frac{m}{n}$ by $\frac{y}{x}$, the result being $\frac{my}{nx}$.

$$\frac{m}{n} \div \frac{x}{y} = \frac{m}{n} \cdot \frac{y}{x} = \frac{my}{nx}.$$

First indicate the work and then cancel if possible, as follows:

$$\frac{p}{q^2} \cdot \frac{3q^3}{p^4} \div \frac{6q}{5p} = \frac{p \cdot 3q^3 \cdot 5p}{q^2\ p^4\ 6q} = \frac{5}{2p^2}.$$

WRITTEN EXERCISE

1. $\frac{a}{bc} \div \frac{a^2}{b^2c^2}$.

2. $\frac{p^2q}{mn} \div \frac{mn}{pq^2}$.

3. $\frac{-a^2bc}{x^2yz} \div \frac{bcd^2}{yz}$.

4. $\frac{a+b}{c} \div \frac{cd^2}{ab}$.

5. $\frac{p^2-q^2}{pq} \div \frac{xy}{m^2}$.

6. $\frac{abc^2}{pqr^2} \div \frac{-bca^2}{p^2qr}$.

7. $\frac{p^2-4q}{4} \div \frac{pq}{8m}$.

8. $\frac{m(m+n)}{x^2y^2} \div \frac{xy}{m^3}$.

9. $\frac{a(a-2b)}{2b} \div \frac{a^3b}{4c}$.

10. $\frac{a^2+b^2}{m^2} \div \frac{m^4}{n^4}$.

11. $\frac{-m^2n}{xy^2} \div \frac{-n^2m}{yx^2}$.

12. $\frac{a^2+2ab+b^2}{ab} \div \frac{ab}{x}$.

13. $\frac{a}{b}\cdot\frac{c}{d}\div\frac{a^2c^2}{b^2d^2}$.

14. $\frac{m^2}{n}\cdot\frac{n^2}{m}\div\frac{m^3}{n^3}$.

15. $\frac{a^2b}{c^2d}\cdot\frac{b^2c}{d^2e}\div\frac{abc}{de}$.

16. $\frac{m^2n}{p^2q}\cdot\frac{n^2p}{q^2m}\div\frac{mnp}{q}$.

17. $\frac{a}{b}\cdot\frac{c}{d}\cdot\frac{e}{f}\div\frac{a^2b^2c^2}{d^2e^2f^2}$.

18. $\frac{x^2y^2z^2}{abc}\cdot\frac{a^2b^2c^2}{xyz}\div\frac{abc}{xyz}$.

19. $\left(\frac{x^2}{y}-\frac{y^2}{x}\right)\div\frac{pq}{xy}$.

20. $\left(\frac{p}{q^2}+\frac{q}{r^2}+\frac{r}{s^2}\right)\div\frac{3}{q^2r^2s^2}$.

21. $\left(\frac{a}{b}+\frac{b}{c}\right)\cdot\frac{a}{bc}\div\frac{1}{b^2c^2}$.

22. $\frac{1+a+a^2}{a^3}\cdot\frac{a}{b^3}\cdot\frac{a}{c^3}\div\frac{-1}{ab^3c^3}$.

23. $\frac{a+b+c}{ab}\ \frac{a^2b}{c}\div\frac{a}{c}$.

24. $\frac{27\,a^2b^2c^2}{17\,m^3n^3}\ \frac{119\,m^2n^2}{18\,a^3b^3}\div\frac{3}{2}$.

25. $\left(\frac{x}{y}-\frac{y}{z}\right)\cdot\frac{y^2z^2}{3}\div\frac{yz}{6}$.

26. $\left(\frac{m}{n}-\frac{n}{m}\right)\cdot\frac{-m^2n^2}{pq}\div\frac{mn}{pq}$.

27. $\frac{a+b-c-d}{abcd}\div\frac{abcd}{4}$.

28. $\left(\frac{p+q}{pq}+\frac{q+r}{qr}\right)\div\frac{1}{pqr}$.

29. $\left(\frac{a}{b}-\frac{c}{d}+\frac{e}{f}\right)\cdot\frac{bdf}{ace}\div\frac{1}{ace}$.

30. $\frac{x^2y}{w^2u}\cdot\frac{-y^2z}{u^2v}\cdot\frac{u^2w}{x^2y}\div\frac{-uvw}{xyz}$.

31. $\frac{x^2-2xy+y^2}{xy}\cdot\frac{-x^3}{y^2z}\div\frac{-x}{y^3z}$.

32. $\frac{-a}{b}\cdot\frac{-b}{c}\cdot\frac{-c}{d}\ \frac{-d}{e}\div\frac{a^2b^2c^2d^2}{2}$.

33. One train travels m miles in h hours; another, d miles in t hours. The first rate is how many times the second?

34. One man earns d dollars in w weeks, and another earns m dollars in t weeks. The weekly wages of the first are how many times those of the second?

LINEAR EQUATIONS

91. Various kinds of equations. — There are many kinds of equations. The equation $x + 2 = 5$ is very simple; $\frac{2}{3}x - 7 = 6$ is more difficult; $x^2 + 7x = 18$ is still more difficult.

92. Linear equation. — An equation like $\frac{2}{3}x - 7 = 6$, in which the unknown quantity has no exponent except 1, is called a *linear equation* or a *simple equation.*

The name *linear* is the more common in advanced work, but both names are used. The expression x^1 means the same as x.

93. Solution of an equation. — To find the value of the unknown quantity is to *solve* the equation.

94. Axiom. — A statement assumed to be true is called an *axiom*

95. Axioms needed in algebra. — The following include the axioms already used (page 4) in solving equations.

Axiom 1. *If equals are added to equals, the sums are equal.*

That is, if $x - 3 = 7$, then $x = 7 + 3$, or 10.

Axiom 2. *If equals are subtracted from equals, the remainders are equal.*

That is, if $x + 3 = 9$, then $x = 9 - 3$, or 6.

Axiom 3. *If equals are multiplied by equals, the products are equal.*

That is, if $\frac{1}{4}x = 5$, then $x = 4 \times 5$, or 20.

Axiom 4. *If equals are divided by equals, the quotients are equal.*

That is, if $7x = 35$, then $x = 35 - 7$, or 5.

Axiom 5 *Quantities which are equal to the same quantity, or to equal quantities, are equal to each other.*

ORAL EXERCISE

1. Solve the equation $x + 2 = 5$; also $x + 2 = b$. Is b considered as known or as unknown? (See page 14, § 16.)

2. Solve the equation $x + 4 = b$; also $x + a = b$.

Solve the following equations, finding the value of x:

3. $x + n = m$.	4. $x - n = m$.	5. $3x = 6$.
6. $3x = a$.	7. $ax = b$.	8. $2ax = 6a^2$.
9. $7x = 14a$.	10. $mx = 3m^3$.	11. $a^2x = 42a^3$.
12. $x + 3a = 7a$.	13. $x - 6a = 14a$.	14. $2x - a = 3a$.

WRITTEN EXERCISE

1. Solve $11x + \frac{3}{4} = \frac{x}{4} + 46$.

2. Solve $7x - 13 = 3x + 27$.

3. Solve $2x + 3x + 7 = \frac{2}{3}x + 33$.

4. $5\frac{x}{2} = 3000$. Find the value of x.

5. Solve $3(\frac{1}{4}x + 200) = 1200$. (First use Axiom 4.)

6. Find the value of x in the equation $5x + 10 = 3x + 32$.

7. Find the number whose fifth and seventh together make 24.

8. Find the number whose half, third, and fourth together make 13.

9. Find the number which added to 4 equals 7 more than half the number.

10. A man saved half of his wages for 5 years. He then had saved $3000. What were his annual wages?

11. A man by saving $200 more than $\frac{1}{4}$ of his wages annually for 3 years, saved $1200. What were his annual wages?

96. How to solve equations. — We have seen that the solution of a linear equation consists in arranging the x's in one member and the known quantities in the other, and dividing by the coefficient of x.

97. Transposition. — The subtracting of terms from both sides of an equation so as to carry them from one side to the other, changing the sign, is called *transposition.*

Thus, in the equation $3 - x = 12 - 4x$, we subtract 3 from both sides, leaving $-x = 12 - 4x - 3$.

We then subtract $-4x$, or, what is the same thing, add $4x$,

and $-x + 4x = 12 - 3$,

or $3x = 9$,

whence $x = 3$.

In this solution we *transposed* the 3 and the $-4x$.

98. Solving by transposition. — Since we have now solved so many equations that we can use the word understandingly, we may say that to solve a linear equation,

Transpose the terms involving x *to the left side, and the known terms to the right, and divide by the coefficient of* x.

Thus, if $\frac{3x}{4} + 5 = \frac{x}{4} + 7$,

then, transposing, $\frac{x}{2} = 2$, or $\frac{1}{2}x = 2$.

The coefficient of x is $\frac{1}{2}$. We may divide by $\frac{1}{2}$, or, what amounts to the same thing, multiply by 2, and

$$x = 4.$$

Check. $\frac{3}{4} \cdot 4 + 5 = \frac{1}{4} \cdot 4 + 7$.

WRITTEN EXERCISE

1. $42x - 1 = x + 81$.
2. $13x - 21 = 4x - 3$.
3. $12x - 6 = 9x + 27$.
4. $15x - 9 = 11x + 39$.
5. $75x - 32 = 16x + 86$.
6. $25x + 8 = 16x + 89$.

7. $\frac{x}{2} - \frac{x}{3} = 7.$

8. $\frac{x}{3} - \frac{x}{4} = 9.$

9. $\frac{2x}{7} + \frac{x}{8} = 69.$

10. $\frac{4x}{5} - \frac{x}{3} = 11.$

11. $\frac{x}{2} + 3 = \frac{x}{8} + 9.$

12. $\frac{5+x}{7} + 3 = 48.$

13. $\frac{9x}{10} - \frac{2x}{3} = 49.$

14. $\frac{x}{2} + \frac{x}{3} + \frac{x}{4} = 13.$

15. $\frac{6x}{7} - \frac{x}{2} - \frac{x}{3} = 1.$

16. $\frac{5x}{7} - \frac{2x}{3} = 135.$

17. $x - 7\% x = 65.1.$

18. $x - 9\% x = 6.37.$

19. $\frac{x+1}{3} + \frac{2}{5} = 31.$

20. $\frac{x-1}{2} + \frac{x+1}{3} = 4.$

21. $x + 10\% x = 605.$

22. $x + 12\% x = 14.56.$

23. $\frac{x-7}{4} - \frac{x+7}{2} = -9.$

24. $\frac{x+7}{5} - \frac{x-1}{2} = 1.$

25. $\frac{x+14}{5} + \frac{17-x}{8} = 5.$

26. $\frac{7x}{9} + \frac{x}{2} - \frac{3x}{10} = 88.$

27. Find a number whose seventh part minus its eleventh part equals 4.

28. Find the number whose third plus 7 equals the number less 3.

29. Find a number which, when subtracted from 83, equals the number less 17.

30. A man lost 32% of his capital and then had $4896. How much had he at first?

31. A man gained 15% on his capital and then had $8625. How much had he at first?

32. A man has 27% of his capital invested in a farm. The farm is worth $1917. How much is his capital?

33. Find a number such that $\frac{5}{7}$ of it less $\frac{3}{8}$ of it equals 19.

34. What sum increased by 1% of itself amounts to \$2585.60?

35. Find a number which increased by 17% of itself, and then decreased by 36, equals 198.

36. A dealer sold some goods for \$2802.40, thus making a profit of 13%. What did the goods cost him?

37. A man lost 30% of his library by fire. He had 630 books left. How many had he before the fire?

38. A collection agency charges 4% for its services in collecting a debt, and remits \$912. How much did it collect?

39. The sum of a certain number, a third of the number, a fourth of the number, less 7, equals $4\frac{1}{12}$. What is the number?

40. A bank charges 0.1% exchange on a draft. The entire cost of draft and exchange is \$1751.75. What is the face of the draft?

41. An agent charges 5% for collecting rents for Mr. Glover. He deducts his commission and remits \$332.50. How much did he collect?

42. Half of the remainder found by subtracting 7 from a certain number equals a fourth of the sum of the number and 7. What is the number?

43. An agent bought a building lot for Mr. Roberts, charging him 3% commission. Mr. Roberts sent him the price of the lot and commission, amounting to \$2626.50. What did the lot cost?

44. The income of a certain store increased $16\frac{2}{3}$% the second year it was open, and the income the third year was 25% more than the second year. The income being \$3500 the third year, what was it the first year?

CHAPTER II

OPERATIONS CONTINUED. FACTORING. PROPORTION. EQUATIONS

MULTIPLICATION

ORAL EXERCISE

1. Multiply x by a; by b. What is the sum?

2. Multiply $m + n$ by a; by b. The sum is the product of $m + n$ by what expression?

3. How can you multiply any polynomial by $a + b$? (Multiply first by a; then by what term? Then what should be done?)

99. Multiplying by a binomial. — Multiplication by a binomial in algebra is much like multiplication by a two-figure number in arithmetic.

Arithmetic:		*Algebra:*	
Multiply 45 by 23.		Multiply $a + 2b$ by $a + b$	
45		$a + 2b$	
23		$a + b$	
135	product by 3	$a^2 + 2ab$	product by a
900	" " 20	$ab + 2b^2$	" " b
1035	" " 23	$a^2 + 3ab + 2b^2$	" " $a + b$

4. Why do you begin at the right to multiply in arithmetic? Why may you begin at the left in algebra? Could you as easily begin at the right in algebra? Try both plans on the blackboard.

ORAL EXERCISE

1. Multiply $4a^6$ by $5a^4$; $-3a^3$ by $6a^2$; $-5x^2$ by $-4x^3$.

2. State the product of $a \cdot -a^2 \cdot 2a \cdot -3a$; also of $3x^2 \cdot -4x^7 \cdot -x$.

3. Tell how you proceed to multiply by $a+b$; by $2a-b$; by $4a^2-3b^3$; by any binomial.

WRITTEN EXERCISE

Multiply in Exs. 1–20:

1. $a+b$ by $a+b$.
2. $x+y$ by $x+y$.
3. $m+n^2$ by $m+n^2$.
4. $2a+b$ by $2a+b$.
5. $a-b$ by $a-b$.
6. $x-y$ by $x-y$.
7. $3m-n$ by $3m-n$.
8. $3-2x^2$ by $3-2x^2$.
9. $a+b$ by $a-b$.
10. $a-b$ by $a+b$.
11. $x-2y$ by $x+2y$.
12. m^2+3n by m^2-3n.
13. $2a+3b$ by $a-2b$.
14. $7x^2y^2+1$ by $2x^2y^2-3$.
15. $6x^2+1$ by $5x^2-3$.
16. $4mn+xy$ by $3mn+4xy$.
17. $a+b+c$ by $a+b$.
18. $x^2+2xy+y^2$ by $x+y$.
19. $m^2-2mn+n^2$ by $m-n$.
20. $4a^2+4ab+b^2$ by $2a+b$.

21. What is the product of 27 and 23? of $a+b$ and $a-b$? Suppose $a=25$ and $b=2$?

22. How many square feet in a square that is 42 ft. on a side? How many square feet in one that is $f+t$ ft. on a side? Suppose $f=40$ and $t=2$?

23. If a man earns $27 a week for 27 weeks, how much does he earn in all? If he earns $t+s$ dollars a week, how much does he earn in $t+s$ weeks?

24. The product $(x+y)(x-y)=x^2-y^2$. Hence write down, without multiplying, the following products: $(10+2)(10-2)$; $12 \cdot 8$; $(20+1)(20-1)$; $21 \cdot 19$.

ORAL EXERCISE

1. State the product of $a + b$ by a; by b. Add them.

2. Write on the board the product of $x + y$ by $x + y$. How are the terms formed from x and y?

100. Squaring a + b, a − b. — Consider the squares of $a + b$ and of $a - b$.

The square of $a + b$.		The square of $a - b$.	
$a + b$		$a - b$	
$a + b$		$a - b$	
$a^2 + ab$	product by a	$a^2 - ab$	product by a
$ab + b^2$	" " b	$- ab + b^2$	" " $-b$
$a^2 + 2ab + b^2$	" " $a+b$	$a^2 - 2ab + b^2$	" " $a-b$

101. Square of any binomial. — It is therefore seen that the first term of the product is the square of the first term of the binomial. The second is twice the product of the two terms (negative when one of them is). The third is the square of the last term of the binomial. Therefore

The square of a binomial equals the sum of the square of the first term, twice the product of the two terms, and the square of the second term.

3. State the squares of $b + c$; of $d + c$; of $2 + a$; of $x^2 + 1$.

4. State the squares of $a - x$; of $x - y$; of $x - 2$; of $x^2 - y^2$.

WRITTEN EXERCISE

Write out the squares of the following without multiplying:

1. $p + q$.	2. $p - q$.	3. $x + t$.
4. $x^2 - t$.	5. $bc - x$.	6. $2b - d$.
7. $ab + c$.	8. $2a + b$.	9. $m^2n + 1$.
10. $m^2n^2 + p$.	11. $xyz - 1$.	12. $p^2qr + 2$.
13. $2a + 3b$.	14. $3a - 2b$.	15. $5xy + 1$.

ORAL EXERCISE

State rapidly the squares of the following:

1. $p + r$.	**2.** $a - 6$.	**3.** $5 + a$.
4. $g - h$.	**5.** $s^2 + t^2$.	**6.** $m + 9$.
7. $x - 7$.	**8.** $d^2 - b^2$.	**9.** $x^3 + y^3$.
10. $xyz + 1$.	**11.** $xyz - 3$.	**12.** $x^2y^2 + z^2$.

WRITTEN EXERCISE

Write out the results without stopping to multiply:

1. $(ab + c)^2$.	**2.** $(3 - y^3)^2$.	**3.** $(y^5 + 1)^2$.
4. $(2 - x^6)^2$.	**5.** $(a + 5c)^2$.	**6.** $4(x + y)^2$.
7. $(x^3 + 2)^2$.	**8.** $(1 - 7q)^2$.	**9.** $(y^3 + x^3)^2$.
10. $(p^3 + 1)^2$.	**11.** $(3a + 2)^2$.	**12.** $(6a - b)^2$.
13. $(x - 11)^2$.	**14** $(x^2y^3 + 1)^2$.	**15.** $(x - 7y)^2$.
16. $(2a - b)^2$.	**17.** $(xy - 2z)^2$.	**18.** $(3 + 5x)^2$.
19. $(a + 7b)^2$.	**20.** $(10 - 3y)^2$.	**21.** $(wxy + z)^2$.
22. $(20x + 1)^2$.	**23.** $(30x - 1)^2$.	**24.** $(3m^2 + 5)^2$.
25 $(xyz + 10)^2$.	**26.** $(abcd + 1)^2$.	**27.** $(11 + 2x)^2$.
28 $100(x + y)^2$.	**29.** $(2x + 2y)^2$.	**30.** $(10x + 10y)^2$.

102. Pictures of squares. — In the figure point to the line that equals $x + y$. Point to the area x^2; to an area xy; to another area xy; to y^2. What then does the square on $x + y$ equal?

y	xy	y	y^2
	x		y
x	x^2	x	xy
	x		y

31. Draw a figure showing the square on $10 + 2$.

32. Draw a figure showing the square on $10 + 1$.

33. Draw a figure showing the square on $2x + y$.

34. Draw a figure showing the square on $3x + x$.

103. Product of a + b and a − b. — Another product frequently met is that of the sum and difference of two quantities.

$$\begin{array}{lll} a + b & \qquad & a - b \\ \underline{a - b} & & \underline{a + b} \\ a^2 + ab & & a^2 - ab \\ \underline{\quad - ab - b^2} & & \underline{\quad\quad ab - b^2} \\ a^2 \quad\quad - b^2 & & a^2 \quad\quad - b^2 \end{array}$$

That is, *the product of the sum and difference of two quantities equals the difference of their squares.*

WRITTEN EXERCISE

Write the following products without stopping to multiply:

1. $(x + 1)(x - 1)$.
2. $(1 + a)(1 - a)$.
3. $(a^2 - 1)(a^2 + 1)$.
4. $(ab - 1)(ab + 1)$.
5. $(abc + 2)(abc - 2)$.
6. $(2x + y^2)(2x - y^2)$.
7. $(1 + mx^2)(1 - mx^2)$.
8. $(p^2 + 4q)(p^2 - 4q)$.
9. $(3 - 4x^3)(3 + 4x^3)$.
10. $(5abc - d)(5abc + d)$.
11. $(10a^2 - 1)(10a^2 + 1)$.
12. $(12a^2 + 11)(12a^2 - 11)$.
13. $(5xyz + 7)(5xyz - 7)$.
14. $(100m + 3)(100m - 3)$.
15. $(12 + 7)(12 - 7)$; $19 \cdot 5$.
16. $(11 - 4)(11 + 4)$; $7 \cdot 15$.
17. $(xyz + 1)^2$.
18. $(7 - 2xy)^2$.
19. $(3x^2 + 4)^2$.
20. $(3 + 2px)^2$.
21. $(3mn - 7)^2$.
22. $(2mn - 3)^2$.
23. $(10 + 3mn)^2$.
24. $(x^2y^2z^2 - 2)^2$.
25. $(5x^2 + 3y)^2$.
26. First write the product of the two binomials; then square the result: $[(a + b)(a - b)]^2$.
27. In the same way write the results of the following: $[(m^2+1)(m^2-1)]^2$; $[(2x-3)(2x+3)]^2$; $[(1-5a)(1+5a)]^2$.

104. The product of two binomials. — The product of two binomials like $x+2$ and $x+5$ is so frequently needed as to require attention. Consider two such cases:

$$\begin{array}{r} x+2 \\ x+5 \\ \hline x^2+2x \quad\quad \\ 5x+10 \\ \hline x^2+7x+10 \end{array} \qquad\qquad \begin{array}{r} x-7 \\ x+3 \\ \hline x^2-7x \quad\quad \\ 3x-21 \\ \hline x^2-4x-21 \end{array}$$

ORAL EXERCISE

1. In the first product how is the 7 (the coefficient of x) formed from the 2 and 5? How is the coefficient of x formed in the second product?

2. How is the last term formed in each of the products? Can you now tell the product of $x+3$ and $x+5$ without actually multiplying?

105. Absolute term. — In the expression $x^2+7x+10$, 10 is called the *absolute term.*

106. $(x+a)(x+b)$. — *The product of* $x+a$ *and* $x+b$ *is* x^2 *plus* $(a+b)x$ *plus* ab.

WRITTEN EXERCISE

Write out the products without stopping to multiply:

1. $(x+7)(x+1)$.
2. $(x+1)(x-1)$.
3. $(a+10)(a-7)$.
4. $(pq+8)(pq+9)$.
5. $(xy-3)(xy-3)$.
6. $(3x+5)(3x-5)$.
7. $(mn-11)(mn+1)$.
8. $(mn+6)(mn+6)$.
9. $(p^2q+7)(p^2q+10)$.
10. $(abc+6)(abc-7)$.
11. $(xyz+20)(xyz-5)$.
12. $(xyz+15)(xyz+2)$.

107. Checks on multiplication. — Because the product of $x + a$ and $x + b$ is $x^2 + (a + b)x + ab$, whatever values may be given to x, a, and b, we may check our work by giving any convenient values to these letters.

For example, to check $(x + 3)(x - 2) = x^2 + x - 6$.

Let $x = 1$. Then $(1 + 3)(1 - 2) = 4 \cdot -1 = -4$, and $1^2 + 1 - 6 = -4$; so the work is probably correct.

Try also $x = 5$. Then $(5 + 3)(5 - 2) = 8 \cdot 3 = 24$, and $5^2 + 5 - 6 = 25 + 5 - 6 = 24$.

108. Ease of checks. — It is usually easier to check the work than to look at an answer in a book.

Work:	*Check:* $x = 1$, $y = 1$.
$2x + 3y$	$= 5$
$4x - 7y$	$= -3$
$8x^2 + 12xy$	-15
$\quad - 14xy - 21y^2$	
$8x^2 - 2xy - 21y^2$	$= -15$

WRITTEN EXERCISE

Multiply and check:

1. $(23x + 21)(x - 17)$.
2. $(21x + 37)(9x - 14)$.
3. $(x^2 + 15y)(x^2 + 14y)$.
4. $(15x^2 + 7)(17x^2 - 8)$.
5. $(17x + 2y)(15x - 6y)$.
6. $(35x + 23y)(15x - 7y)$.
7. $(11y - 2x)(2y - 11x)$.
8. $(31xyz + 2)(17xyz - 7)$.
9. $(15a^3b^2c + 7)(9a^3b^2c - 8)$.
10. $(27x^2y^2 + 1)(30x^2y^2 - 3)$.
11. $(32x + 3y)(-41x + 8y)$.
12. $(231a + 147)(329a - 176)$.
13. $(321a^2 + 17b)(151a^2 - 5b)$.

14. $(a+b)(a^2-ab+b^2)$.
15. $(a-b)(a^2+ab+b^2)$.
16. $(a+b)(a^2+2ab+b^2)$.
17. $(a-b)(a^2-2ab+b^2)$.
18. $(a+b)(a-b)(a^2+b^2)$.
19. $(81abc-1)(91abc+17)$.
20. $(62abc+1)(78abc+3)$.
21. $(2x-3)(4x^2-7x+2)$.
22. $(4x-1)(8x^2-2x+3)$.
23. $(42a-37b)(51a-17b)$
24. $(111a^2x+21)(97a^2x+3)$.
25. $(4x^2-4xy+y^2)(2x-y)$.
26. $(9x^2+6xy+y^2)(3x+y)$.
27. $(a+3b)(a+4b)(a+5b)$.
28. $(172x^2y^2+z^2)(172x^2y^2-z^2)$.
29. $(42p^2q^2r^2+s^2)(17p^2q^2r^2-s^2)$.
30. $(a+b)(a+b)(a+b)(a+b)$.
31. $(7x+2y)(9x^2+4xy-6y^2)$.
32. $(a^2b^2c^2-2abcd+d^2)(abc-d)$.
33. $(ax^2+by^2)(ax^4+abx^2y^2+by^4)$.
34. $(a+b)(a^3+3a^2b+3ab^2+b^3)$.
35. $(a^3b^2c+1)(a^6b^4c^2+2a^3b^2c+1)$.
36. $(8x-3y)(6x^2-7xy+50y^2)$.
37. $(x-y)(x^3-3x^2y+3xy^2-y^3)$.
38. $(21x^2-1)(152x^4+73x^2-41)$.
39. $(-27x^2y^2-z^2)(-21x^2y^2-z^2)$.
40. $(2x^3-7y)(9x^6+15x^3y+8y^2)$.
41. $(6a+7b)(36a^2+84ab+49b^2)$.

109. Illustrative problem. — At what rate will \$320 produce \$19.20 interest in two years?

1. If the rate for 1 yr. is $r\%$, for 2 yr. it is $2r\%$.
2. Since $2r\%$ of \$320 = \$19.20,

$$r\% = \tfrac{1}{2} \text{ of } \frac{\$19.20}{\$320} = 3\%.$$

WRITTEN EXERCISE

1. What per cent of \$75 is \$3.75?

2. What per cent of \$15.50 is \$0.62?

3. If $x\%$ of \$156 is \$4.68, find the value of x.

4. If $x\%$ of \$730 is \$65.70, find the value of x.

5. What sum increased by 6% of itself equals \$1007?

6. On what sum is \$11.25 the interest for 1 year at $4\frac{1}{2}\%$? at 5%? at 3%? at 2%?

7. At what rate of interest will \$450 produce \$15.75 in 1 year? in 2 years? in 9 months?

8. At what rate will \$240 produce \$28.80 interest in 2 years? in 3 years? in 4 years?

9. How long will it take \$350 and interest to amount to \$386.75 at $3\frac{1}{2}\%$? at 5%? at 4%?

10. At what rate will \$260 and interest amount to \$282.75 in $3\frac{1}{2}$ years? in 5 years? in $2\frac{1}{2}$ years?

11. What sum will amount, with interest, to \$381.50 in 2 years at $4\frac{1}{2}\%$? at 5%? at 6%? at 2%?

12. A man lends \$250 for a year at a certain per cent, and the next year he lends the \$250 and the first year's interest at the same rate. The sum of principal and interest at the end of the first year is \$260. What is the rate? What is the sum at the end of the second year?

FACTORING

ORAL EXERCISE

1. What are the factors of 15? of ab? of 25? of x^2?

2. Name two factors of 12. Are they prime factors? If not, state the prime factors of 12.

3. What is the product of $a+b$ and $a+b$? What are the factors of $a^2+2ab+b^2$? of $x^2+2xy+y^2$?

4. What is the product of $x-y$ and $x-y$? What are the factors of $x^2-2xy+y^2$? of $m^2-2mn+n^2$?

5. What is the product of $m+n$ and $m-n$? What are the factors of m^2-n^2? of p^2-q^2? of $4-x^2$?

6. What is the product of $x+2$ and $x+3$? What are the factors of x^2+5x+6? of $a^2b^2+5ab+6$?

110. How to factor expressions. — *Factoring always depends upon remembering types of products.*

111. Monomial factors. — Because we remember that $x(y+z)=xy+xz$, it follows that the factors of $xy+xz$ are x and $y+z$. That is,

$$\mathbf{xy + xz = x(y + z).}$$

WRITTEN EXERCISE

Factor the following:

1. $xa+xb$.
2. $am+mb$.
3. x^2+2xy.
4. $pq-qr$.
5. x^2y+y^2x.
6. p^2-3pq.
7. $abc+bcd$.
8. ax^2+axy.
9. $ab+5bc$.
10. $3pq^2+6q^3$.
11. mx^2-nxy.
12. $c^4+3c^2d^2$.
13. $2x^2y+6y^2z$.
14. $abxy+wxyz$.
15. $mxy+nxz+qwx$.
16. $32xy^2+8xyz+4xyz^2$.

ORAL EXERCISE

1. Reduce to lowest terms these fractions: $\frac{4}{12}, \frac{5}{15}, \frac{16}{24}, \frac{ab}{ac}$.

2. Also the following: $\frac{p^2q}{pq^2}, \frac{a(a+b)}{b(a+b)}, \frac{x(x+1)}{y(x+1)}, \frac{ab(m+n)}{xyzab}$.

3. How do you ordinarily reduce fractions to lowest terms? Give three illustrations.

112. Uses of factoring. — One of the chief applications of factoring is in the reduction of fractions to lowest terms, so that the fractions may be used more easily.

For example, to reduce the fraction $\frac{x^2+3x}{x^2+4x}$ to lowest terms.

$$\frac{x^2+3x}{x^2+4x} = \frac{x(x+3)}{x(x+4)} = \frac{x+3}{x+4}.$$

Also, to reduce the fraction $\frac{x^2-7x}{xy-7y}$ to lowest terms.

$$\frac{x^2-7x}{xy-7y} = \frac{x(x-7)}{y(x-7)} = \frac{x}{y},$$

by factoring and then canceling $x-7$ from both terms.

WRITTEN EXERCISE

Reduce to lowest terms the following fractions:

1. $\frac{m^2+m}{m^2-m}$.
2. $\frac{x+x^2}{x^2-x}$.
3. $\frac{a-ab}{a^2+a}$.
4. $\frac{x^2+xy}{xy+zx}$.
5. $\frac{m^2+mn}{n^2+mn}$.
6. $\frac{4x^2+6x}{2xy+8x}$.
7. $\frac{2pq+q^2}{2pr+rq}$.
8. $\frac{3mx+x}{y+3my}$.
9. $\frac{81-27y}{63+18y}$.
10. $\frac{a^2-7a}{abc-7bc}$.
11. $\frac{xy-3y^2}{zx-3yz}$.
12. $\frac{pq^2r-p^2qs}{3p^2q^2+pqr}$.

ORAL EXERCISE

1. What is the square of $a+b$? of $a-b$?

2. Square $x+y$; $x-y$; $2x+1$; $1-3y$; $ab+cd$.

3. What quantity squared equals $a^2+2ab+b^2$? $x^2-2xy+y^2$? $4x^2+4x+1$? x^4+2x^2+1?

4. What are the two equal factors of m^2-2m+1? of $x^2y^2+2xy+1$? $a^2+4ab+4b^2$? x^4-2x^2+1?

113. Squares of binomials. — Because we know that $(x+y)^2 = x^2+2xy+y^2$, and $(x-y)^2 = x^2-2xy+y^2$, it follows that the factors of these trinomials are known:

$$\mathbf{x^2 + 2xy + y^2 = (x+y)^2.}$$
$$\mathbf{x^2 - 2xy + y^2 = (x-y)^2.}$$

For example, the factors of $49x^2+14x+1$ are $7x+1$ and $7x+1$. For

$$49x^2+14x+1 = (7x)^2+2(7x)+1$$
$$= (7x+1)^2.$$

So the factors of $9x^2-12xy+4y^2$ are $3x-2y$ and $3x-2y$. For

$$9x^2-12xy+4y^2 = (3x)^2-2(3x)(2y)+(2y)^2$$
$$= (3x-2y)^2.$$

WRITTEN EXERCISE

Factor the following:

1. $x^2-2xm+m^2$.
2. $x^2+2xn+n^2$.
3. x^4+2x^2+1.
4. $1-2x^2+x^4$.
5. $4x^2+4x+1$.
6. $1-4x^2+4x^4$.
7. $9x^2+6x+1$.
8. $16x^2-8x+1$.
9. $x^2y^2+2xyz+z^2$.
10. $a^2b^2-2abcd+c^2d^2$.
11. $p^4q^2+20p^2q+100$.
12. $a^2b^2c^2+121+22abc$.
13. $9a^4+42a^2b^2+49b^4$.
14. $36x^4+60x^2y^2+25y^4$.

ORAL EXERCISE

1. What is the product of $a + b$ and $a - b$? What are the factors of $a^2 - b^2$? of $x^2 - y^2$? of $m^2 - n^2$?

2. What is the product of $2x + 1$ and $2x - 1$? What are the factors of $4x^2 - 1$? of $4m^2 - 1$? of $1 - 4a^2$?

3. What are the factors of the difference of two squares?

114. Difference of squares. — Because we know that $(x + y)(x - y) = x^2 - y^2$, it follows that *the factors of the difference of the squares of two quantities are the sum and difference of the quantities.*

$$\mathbf{x^2 - y^2 = (x + y)(x - y).}$$

For example, the factors of $25x^4 - 121y^2$ are $5x^2 + 11y$ and $5x^2 - 11y$. For

$$\begin{aligned} 25x^4 - 121y^2 &= (5x^2)^2 - (11y)^2 \\ &= (5x^2 + 11y)(5x^2 - 11y). \end{aligned}$$

Similarly, $(x - y)^2 - 4 = (x - y + 2)(x - y - 2)$.

WRITTEN EXERCISE

Factor the following:

1. $p^2 - 4q^2$.
2. $9a^2 - 4b^2$.
3. $16x^4 - y^2$.
4. $64m^4n^2 - 1$.
5. $49 - 121m^2$.
6. $1 - 100a^2b^2c^2$.
7. $(a + b)^2 - c^2$.
8. $2ax + a^2 + x^2$.
9. $36a^2 - 25b^2$.
10. $144m^2 - 25n^2$.
11. $36p^2q^4 - 121$.
12. $81 - 25a^2b^2c^2d^2$.
13. $a^4 + 2a^2b + b^2$.
14. $25a^2b^2 - 25b^2c^2$.
15. $xa^2 + b^2x + cxy$.
16. $4m^2n + n^2 + 4m^4$.
17. $64a^4b^2 - 81c^2d^4$.
18. $a^2 - 4ab + 4b^2 - 16$.
19. $a^2 + 2ab + b^2 - 4$.
20. $1 + 6(x + y) + 9(x + y)^2$.

ORAL EXERCISE

1. What is the product of $x + 1$ and $x + 2$? What are the factors of $x^2 + 3x + 2$?

2. What is the product of $x - 3$ and $x + 4$? What are the factors of $x^2 + x - 12$? of $x^2 + xy - 12y^2$?

3. Of what expression are $x + 5$ and $x + 4$ the factors? also $x - 7$ and $x + 4$? also $x + 1$ and $x - 8$?

115. The product of two binomials. — Because we know that $(x + a)(x + b) = x^2 + (a + b)x + ab$, it follows that we can often tell the factors of expressions like

$$\begin{array}{lr} x\ + & b \\ x\ + & a \\ \hline x^2 + & bx \\ & ax + ab \\ \hline x^2 + (a+b)x + ab & \end{array}$$

$$x^2 + 7x - 18 \text{ and } x^2 + 5x + 6.$$

For example, the factors of $x^2 + 7x - 18$ are $(x + 9)(x - 2)$.

For $$x^2 + 7x - 18 = x^2 + (9 - 2)x + 9(-2)$$
$$= (x + 9)(x - 2).$$

That is, 9 and -2 are two numbers which *added make the coefficient of* x, *and multiplied make the third term.*

WRITTEN EXERCISE

Factor the expressions in Exs. 1–14:

1. $x^2 + x - 42$.
2. $x^2 + 7x + 10$.
3. $x^2 - 5x + 6$.
4. $a^2 - 7a - 18$.
5. $p^2 + 18p + 81$.
6. $x^4 + 7x^2 - 18$.
7. $m^2 + 12m + 35$.
8. $p^2 - 20p + 100$.
9. $p^2q^2 + 13pq + 36$.
10. $a^2b^2 + 6ab - 40$.
11. $x^2y^2z^2 + 11xyz + 24$.
12. $m^2n^2 + 17mn + 72$.
13. $mn^2 - mp^2 - q^2m$.
14. $(a + b)^2 - (c + d)^2$.
15. Of what expression are $ab^3 - 4cd^2$ and $ab^3 + 17cd^2$ the factors?

ORAL EXERCISE

Reduce to lowest terms:

1. $\frac{a+2b}{a^2-4b^2}$.

2. $\frac{p-q}{p^2-q^2}$.

3. $\frac{m+n}{m^2-n^2}$.

4. $\frac{m(m-1)}{m^2-1}$.

5. $\frac{a(b+c)}{b^2-c^2}$.

6. $\frac{x+y}{x^2+2xy+y^2}$.

7. $\frac{a^2+2ab+b^2}{a^2-b^2}$.

8. $\frac{(x-y)^2}{x^2-2xy+y^2}$.

9. $\frac{x-y}{x^2-2xy+y^2}$.

10. $\frac{x^2-y^2}{x^2-2xy+y^2}$.

116. Reduction of fractions. — As usual, in reducing fractions to lowest terms, *first factor, then cancel common factors.*

WRITTEN EXERCISE

Reduce to lowest terms:

1. $\frac{a^2+2ab+b^2}{a^2b+ab^2}$.

2. $\frac{a^2+a-12}{a^2-a-20}$.

3. $\frac{1+x^2}{x^4+2x^2+1}$.

4. $\frac{a^2+2a-8}{a^2+6a+8}$.

5. $\frac{x^2-5x+6}{x^2-2x-3}$.

6. $\frac{(a+b)^2-1}{a+b+1}$.

7. $\frac{p^2-5p+6}{p^2-8p+12}$.

8. $\frac{x^8+2x^4+1}{x^4y^4+y^4}$.

9. $\frac{a^2x+ya^2+a^2z}{(x+y)c^2+c^2z}$.

10. $\frac{x^2-3x+2}{x^2-22x+40}$.

11. $\frac{1-15x+56x^2}{1-17x+72x^2}$.

12. $\frac{4m^2-n^2}{4m^2+n^2+4mn}$.

WRITTEN EXERCISE

1. Solve the equation $3x + 9 = 5x - 65$.

2. What sum increased by 3% of itself amounts to $785.89?

3. What sum decreased by 7% of itself is reduced to $709.59?

4. The cost of a draft, less the discount at 0.1%, is $9740.25. What is the face?

5. The cost of a draft, including the premium at 0.1%, is $10,760.75. What is the face?

6. A man offered $7790 for a farm, which was 5% less than the asking price. What was the asking price?

7. A man pays an agent $7725 for buying a house, which includes the agent's commission of $225. What was the per cent of commission?

8. A man pays an agent $5610 for buying a house, which includes the agent's commission of 2%. What did the agent pay for the house?

9. A man bought two horses at the same price each. He sold the two for $198.90, gaining 11% on one and 10% on the other. What did he pay for them?

10. A man bought two horses at the same price each. He sold the two for $163 20, gaining 10% on one and losing 6% on the other. What did he pay for them?

11. A man increases his original capital by 7%, and the next year he decreases what he then had by 10%. He then had $8667. What was his original capital?

12. A mine increased its income 11% in one year, and the next year it increased this new income 10%. The income then amounted to $335,775 a year. What was its income at first?

DIVISION

ORAL EXERCISE

1. Divide $4a^2b$ by a^2; by b; by $2a$.

2. Divide $xa + xb$ by x; by $a + b$. Divide $m^2 - mn$ by m.

3. What are the factors of $a^2 - 2ab + b^2$? Divide $a^2 - 2ab + b^2$ by $a - b$.

4. Divide $a^2 - b^2$ by $a - b$; by $a + b$. Divide $m^2 + 2m + 1$ by $m + 1$. Divide $x^4 - 1$ by $x^2 - 1$.

117. Dividing by binomials. — We have just seen some examples of dividing by binomials. Dividing in algebra is quite like dividing in arithmetic.

Arithmetic:

Divide 3772 by 46.

$$\begin{array}{r@{\,}l} & 82 \\ 46\overline{)\,3772} & \\ 3680 & = 80 \text{ times } 46 \\ \hline 92 & \\ 92 & = 2 \text{ " " } \\ \hline \end{array}$$

Algebra:

Divide $2a^2 + 7ab + 3b^2$ by $a + 3b$.

$$\begin{array}{r} 2a + b \\ a + 3b\overline{)\,2a^2 + 7ab + 3b^2} \\ 2a^2 + 6ab \qquad \\ \hline ab + 3b^2 \\ ab + 3b^2 \\ \hline \end{array}$$

Check:

$a = 1$, $b = 1$.

$12 - 4 = 3$.

118. The work is easier in algebra, however, since we need, after once arranging both dividend and divisor in the same order as to some letter, to divide only the first term of the dividend by the first term of the divisor to find the first term of the quotient.

Here $2a^2 \div a = 2a$, subtracting $2a(a + 3b)$, there remains $ab + 3b^2$ to be divided. Then $ab \div a = b$, subtracting $b(a + 3b)$, there remains nothing.

Therefore $2a^2 + 7ab + 3b^2 = (2a + b)(a + 3b)$ In other words, the quotient is $2a + b$

119. We check, as in multiplication, by putting numbers for letters, as $a = 1$, $b = 1$.

WRITTEN EXERCISE

Divide in Exs. 1–15:

1. $x^2 + 2xy + y^2$ by $x + y$.
2. $2m^2 - mn - 3n^2$ by $m + n$.
3. $3x^2 + xy - 2y^2$ by $3x - 2y$.
4. $6p^2 + pq - 2q^2$ by $3p + 2q$.
5. $42x^2 - 13xy + y^2$ by $7x - y$.
6. $a^3 + 3a^2b + 3ab^2 + b^3$ by $a + b$.
7. $10a^2 - 29ab + 10b^2$ by $2a - 5b$.
8. $20p^2q^2r^2 + 41pqr + 2$ by $pqr + 2$.
9. $40x^2y^2 + 58xy - 21$ by $10xy - 3$.
10. $30m^4 - 229m^2 + 30$ by $15m^2 - 2$.
11. $3 - 47xy + 170x^2y^2$ by $1 - 10xy$.
12. $32x^2y^2 + 46xyz - 3z^2$ by $16xy - z$.
13. $1 - 18pqr + 77p^2q^2r^2$ by $1 - 11pqr$.
14. $50x^4y^4 + 77x^2y^2z^2 + 3z^4$ by $25x^2y^2 + z^2$.
15. $32x^4 - 50x^2 + 3$ by $16x^2 - 1$; also by $4x + 1$.

16. If a body moves uniformly $4x^2 + 3xy - y^2$ feet in $4x - y$ seconds, how far does it move per second?

17. If $3x + 7$ pounds on a lever will raise $12x^2 + 19x - 21$ pounds, how much will 1 pound raise? How much will $x + 2$ pounds raise?

18. If $x + y$ articles cost $3x^2 + 2xy - y^2$ dollars, how much does each cost? Supposing $x = 1$ and $y = 1$, what is the result?

19. If there are $108x^3 + 144x^2y + 36xy^2$ words in a book of $36x$ pages, how many words are there, on an average, to each page? If there are $3x + y$ lines to a page, how many words, on an average, to a line?

ORAL EXERCISE

1. Divide by x : $ax + bx$, $ax + b$.
2. Divide by $x + y$: $(x + y)^2$, $(x + y)^2 + m$.
3. Divide by $x - y$: $x^2 - y^2$, $x^2 - 2xy + y^2$, $x - y - a$.
4. Divide by $a + 1$: $(a + 1)^2$, $(a + 1)^2 - x$, $(a + 1)^3 - x^2$.

120. Remainders in division. — In dividing by binomials remainders are treated in the same way as in dividing by monomials. That is, they lead to fractions in the quotient.

For example, divide $2xy + x^2 + 3y^2$ by $x + y$. Rearranging the dividend and dividing in the usual way, there is a remainder of $2y^2$. Therefore the quotient is

$$x + y + \frac{2y^2}{x + y}.$$

$$\begin{array}{r|l} & x + y + \dfrac{2y^2}{x+y} \\ x + y & x^2 + 2xy + 3y^2 \\ & \underline{x^2 + xy} \\ & \quad xy + 3y^2 \\ & \quad \underline{xy + y^2} \\ & \quad 2y^2 \end{array}$$

WRITTEN EXERCISE

Divide in Exs. 1–4:

1. $x^2 - 2xy + 7y^2$ by $x - y$; by $x + y$.
2. $m^3 + 3m^2 - 3m$ by $m + 1$; by $m - 1$.
3. $x^3 + 3x^2y + 3xy^2 - 6y^3$ by $x + y$; by $x - y$.
4. $6a^3 - 17a^2b + 11ab^2 + 5b^3$ by $2a - b$; by $2a + b$.

Rearrange the dividend and divide in Exs. 5–9:

5. $17ab + 21a^2 + 5b^2$ by $7a + b$; by $7a - b$.
6. $37pq + 6p^2 + 7q^2$ by $6p + q$; by $6p - q$.
7. $17xy - 13y^2 + 6x^2$ by $3x - 2y$; by $3x + 2y$.
8. $6m^2 + 18mn - 12n^2$ by $2m + 7n$; by $2m + 5n$.
9. $13mn + 3n^2 + 20m^2$ by $5m + 2n$; by $5m - 2n$.

121. Arranging terms. — It often happens that in attempting to arrange the terms of the dividend in the same order as those of the divisor, certain powers will be missing. In that case zeros may be inserted if desired.

For example, to divide $x^3 - y^3$ by $x - y$ either of these forms may be taken. After a little the second one will naturally come to be used.

$$
\begin{array}{r|l}
 & x^2 + xy + y^2 \\
\hline
x - y & x^3 + 0 + 0 - y^3 \\
 & \underline{x^3 - x^2y} \\
 & x^2y + 0 \\
 & \underline{x^2y - xy^2} \\
 & xy^2 - y^3 \\
 & \underline{xy^2 - y^3}
\end{array}
\qquad
\begin{array}{r|l}
 & x^2 + xy + y^2 \\
\hline
x - y & x^3 - y^3 \\
 & \underline{x^3 - x^2y} \\
 & x^2y - y^3 \\
 & \underline{x^2y - xy^2} \\
 & xy^2 - y^3 \\
 & \underline{xy^2 - y^3}
\end{array}
$$

WRITTEN EXERCISE

Divide in Exs. 1–6:

1. $x^3 + y^3$ by $x + y$.
2. $a^5 - b^5$ by $a - b$.
3. $x^4 - y^4$ by $x + y$.
4. $8x^3 - 1$ by $2x - 1$.
5. $x^5 + y^5$ by $x + y$.
6. $1 + 27a^3$ by $1 + 3a$.

In Exs. 7–10 one factor is given in parentheses; required the other factor:

7. $a^3 + 3a^2 + 5a + 3$, $(a + 1)$.
8. $1 - 64x^3y^3$, $(1 - 4xy)$.
9. $32m^5 + 1$, $(2m + 1)$.
10. $a^3b^3c^3 - d^3$, $(abc - d)$.

Since a^4 is equal to $a^2 \cdot a^2$, $a^4 - b^4$ may be thought of as the difference of the squares of a^2 and b^2. Hence

$$a^4 - b^4 = (a^2 + b^2)(a^2 - b^2) = (a^2 + b^2)(a + b)(a - b).$$

Factor the following:

11. $x^4 - y^4$.
12. $16m^4 - n^4$.
13. $16p^4 - 1$.
14. $81x^4 - 16y^4$.
15. $x^4y^4 - m^4n^4$.
16. $81p^4q^4r^4 - 1$.

FRACTIONS

ORAL EXERCISE

1. Reduce to lowest terms: $\frac{a(a+b)}{b(a+b)}, \frac{a(b+c)}{a(c+d)}$.

2. Reduce $\frac{1}{x+y}$ to a fraction with the denominator $(x+y)^2$; with the denominator x^2-y^2.

122. Polynomial denominators. — The method of reducing fractions is the same for polynomial as for monomial denominators.

For example, to reduce $\frac{2x+1}{x-2}$ to a fraction with the denominator x^2-5x+6.

1. $(x^2-5x+6)\div(x-2)=x-3$.
2. Therefore $x-2$ must be multiplied by $x-3$.
3. Therefore both terms must be multiplied by $x-3$ (§ 73).
4. Therefore $\frac{2x+1}{x-2}=\frac{(x-3)(2x+1)}{(x-3)(x-2)}=\frac{2x^2-5x-3}{x^2-5x+6}$.

123. Changing Signs. — It should also be remembered that, since we may multiply the terms by -1, *we may change the signs in both terms.*

For example, $$\frac{x-1}{2-x}=\frac{1-x}{x-2}.$$

WRITTEN EXERCISE

Reduce to fractions with the denominator $x^2-9x+14$:

1. $\frac{x+2}{x-2}$. 2. $\frac{x-3}{x-7}$. 3. $\frac{x^3+6x^2}{14-x(9-x)}$. 4. $\frac{x-1}{2-x}$.

With the denominator $6a^3-29a^2+46a-24$:

5. $\frac{a+2}{a-2}$. 6. $\frac{a-7}{2a-3}$. 7. $\frac{3a+1}{3a-4}$. 8. $\frac{a+1}{3-2a}$.

ORAL EXERCISE

1. Express as mixed numbers: $\frac{3}{2}$, $\frac{5}{3}$, $\frac{9}{7}$.

2. Reduce to mixed expressions: $\dfrac{a+1}{a}$, $\dfrac{ab+ac+d}{a}$.

3. Express in fractional form: $2\frac{1}{3}$, $a+\dfrac{1}{b}$, $x+\dfrac{1}{x+y}$.

124. Mixed quantities reduced to fractions. — Required to express $a-b-\dfrac{a-b^2}{a+b}$ as a fraction.

1. Since $$1=\frac{a+b}{a+b},$$

2. Therefore $$a-b=\frac{(a-b)(a+b)}{a+b}=\frac{a^2-b^2}{a+b}.$$

3. Therefore $$a-b-\frac{a-b^2}{a+b}=\frac{a^2-b^2}{a+b}-\frac{a-b^2}{a+b}$$

4. $$=\frac{a^2-b^2-(a-b^2)}{a+b}$$

5. $$=\frac{a^2-b^2-a+b^2}{a+b}=\frac{a^2-a}{a+b}.$$

WRITTEN EXERCISE

Express in fractional form:

1. $2\frac{1}{2}$.
2. $3\frac{2}{3}$.
3. $a+\dfrac{1}{b}$.
4. $m-\dfrac{1}{m}$.
5. $x+y+\frac{1}{3}$.
6. $x^2+\dfrac{x^3}{x+y}$.
7. $a+\dfrac{b}{xy}$.
8. $a+\dfrac{m}{m+n}$.
9. $x-y-\dfrac{xy}{y}$.
10. $m+1+\dfrac{1}{m}$.
11. $m-1-\dfrac{1}{m}$.
12. $a+b-\dfrac{a^2+b^2}{a-b}$.

13. $a + b^2 - \frac{c}{a-b}$.

14. $a^3 + a^2 + 1 - \frac{1}{a}$.

15. $x^2 - 1 + \frac{1}{x+1}$.

16. $a^2 + 1 - \frac{a^3+1}{a-1}$.

17. $x^2 + x - 3 + \frac{2}{x}$.

18. $p^2 + q + \frac{q^3 - p^3}{p+q}$.

19. $2a - 3 - \frac{a+2}{a-2}$.

20. $p^3 - p + 7 - \frac{8}{3p^2}$.

21. $x^4 + x^2 - \frac{1}{x^2} + 3$.

22. $a^2 + a + 1 - \frac{1}{a} + a^3$.

23. $p^2 - q^2 + \frac{p^3 - 1}{p+q}$.

24. $3a^4 - 2a^3 - \frac{a^6+1}{a^2+1}$.

25. $x^2 + y^2 + \frac{x^3 - y^3}{x+y}$.

26. $3m + 7 - \frac{6m+21}{2m+3}$.

27. $2x - 3 + \frac{4x^2}{4x+7}$.

28. $ax + b - \frac{2a^2x^2 + b^2}{ax - b}$.

29. $m + n - \frac{3m^2 - 2n^2}{m-n}$.

30. $mx + ny - \frac{x + 2n^2y^2}{2ny}$.

31. $a^2 + b^2 + c^2 + \frac{a^3 - ab^2}{a+1}$.

32. $a^2 + 3ab - \frac{a^3 + 3ab^2}{a-b}$.

33. $14a^2b^2 + 3ab - \frac{a^2b^2 - 2}{1 - ab}$.

34. $15m^3 - 7m^2 - \frac{2 - m^5}{m^2 - 1}$.

35. $8x^2y^2 - 3xy - \frac{xy - 1}{xy + 1}$.

36. $m^3 + m^2 + m - \frac{m-n}{m+n}$.

37. $x^2 + 2xy + y^2 + \frac{x^3 - y^3}{x+y}$.

38. $32x^2 + 7x - 3 + \frac{x^2 - 2}{x+1}$.

39. $2p + 3q + \frac{2p^2 - 5pq}{p+q}$.

40. $2a - b + 3c + \frac{ab - 2a^2}{a+b}$.

41. $3m + 5n + \frac{20n^2 - 9m^2}{3m - 5n}$.

42. $3a - b + 2c + \frac{2bc - 3a^2}{a-b}$.

43. $32a^2 - 7b^2 + 6c^2 + \frac{37}{111}$.

44. $9\frac{1}{3}a^2 - 6\frac{2}{3}b^2 + 3\frac{1}{3}c^2 - 6\frac{2}{3}d^2$.

ORAL EXERCISE

Reduce to entire or mixed quantities:

1. $\frac{32a^2b}{16b}$. 2. $\frac{48m^3n}{12mn}$. 3. $\frac{x^2-9y^2}{x+3y}$.

4. $\frac{(a+b)^3}{a+b}$. 5. $\frac{4a^2-b^2}{2a-b}$. 6. $\frac{75(x^2-y^2)}{25(x-y)}$.

7. $\frac{x^2-2xy+y^2}{x-y}$. 8. $\frac{a+(x+y)^2}{a}$. 9. $\frac{(a+b)^2+m}{a+b}$.

125. Illustrative problem. — Reduce $\frac{x^2+3x+9}{x+2}$ to an entire or a mixed quantity.

Since a fraction is an expression of division,

$$\frac{x^2+3x+9}{x+2} = (x^2+3x+9) \div (x+2) = x+1+\frac{7}{x+2}.$$

WRITTEN EXERCISE

Reduce to entire or mixed quantities:

1. $\frac{x^3-y^3}{x+y}$. 2. $\frac{a^4+b^4}{a+b}$.

3 $\frac{x^2+7x+8}{x+1}$. 4. $\frac{x^2-3x-4}{x-2}$.

5. $\frac{x^2+3x+y^2}{x+y}$. 6. $\frac{x^2-9x+6}{x-4}$.

7. $\frac{x^2-8x+11}{x+7}$. 8. $\frac{x^2y^2-4xy+10}{xy-1}$.

9. $\frac{a^3+a^2+a+1}{a+1}$. 10. $\frac{p^2q^2-3pq-88}{pq-11}$.

11. $\frac{x^2y^2+11xy+28}{xy+7}$. 12. $\frac{m^2n^2p^2-23mnp+60}{mnp-20}$.

13. $\frac{a^3-b^3}{a-b}$ and $\frac{a^3+b^3}{a+b}$. 14. $\frac{x^3+3x^2y+3xy^2+y^3}{x+y}$.

ORAL EXERCISE

Reduce to fractions having the lowest common denominator:

1. $\frac{1}{2}, \frac{1}{3}$.
2. $\frac{3}{4}, \frac{1}{8}$.
3. $\frac{5}{6}, \frac{1}{4}$.
4. $\frac{1}{a}, \frac{1}{b}$.
5. $\frac{a}{b}, \frac{c}{d}$.
6. $\frac{a}{bc}, \frac{b}{cd}$.

126. Illustrative problem. — Reduce $\frac{a}{bc}$ and $\frac{x}{cp + cq}$ to fractions having the lowest common denominator.

1. Since the factors of the denominators are b, c, $p + q$, the l.c.d. is $bc(p + q)$.

2. Multiplying the terms of $\frac{a}{bc}$ by $p + q$, we have $\frac{a(p + q)}{bc(p + q)}$.

3. Multiplying the terms of $\frac{x}{cp + cq}$ by b, we have $\frac{bx}{bc(p + q)}$.

WRITTEN EXERCISE

Reduce to fractions having the lowest common denominator:

1. $\frac{1}{a}, \frac{a}{a + b}$.
2. $\frac{1}{m}, \frac{m}{m + 1}$.
3. $\frac{2pr}{2pr - r^2}, \frac{p}{r^2}$.
4. $\frac{a}{xy}, \frac{b}{y^2 + y}$.
5. $\frac{a^2 + 1}{a^3 - a}, \frac{1}{a^2}$.
6. $\frac{m^2 + 1}{m^2 + m}, \frac{1}{m^3}$.
7. $\frac{a}{bc}, \frac{b}{ac + c^2}$.
8. $\frac{m}{n^2}, \frac{1}{mn + n^2}$.
9. $\frac{x}{yz}, \frac{z}{y(x + y)}$.
10. $\frac{p^2 + 4}{p^2 - 4p}, \frac{q}{p}$.
11. $\frac{x}{p^3q^2}, \frac{y}{pq - p^2q^2}$.
12. $\frac{p}{q^2r^2}, \frac{1}{qr + pqr}$.
13. $\frac{x^2 - 2xy + y^2}{x^2 + xy}, \frac{x^2 + 2xy + y^2}{xy}$.
14. $\frac{m - 1}{m}, \frac{m}{m - 1}$.
15. Solve the equation $14x - 175 = 17x - 241$.

ORAL EXERCISE

Reduce to lowest terms:

1. $\frac{3xyz}{6wxy}$.
2. $\frac{2a+b}{4a^2-b^2}$.
3. $\frac{a+b}{a^2-b^2}$.
4. $\frac{x+y}{x^2+2xy+y^2}$.
5. $\frac{4(a+b)}{8(a+b)^2}$.
6. $\frac{m+1}{m^2-1}$.
7. $\frac{xyz(x+y+z)^2}{x^2y^2z^2(x+y+z)}$.
8. $\frac{x-y}{x^2-2xy+y^2}$.
9. $\frac{72(x+y)^3}{36(x+y)^2}$.

127. Illustrative problem. — Reduce the fraction

$$\frac{x^2+8x+15}{x^2+10x+21}$$

to lowest terms.

$$\frac{x^2+8x+15}{x^2+10x+21} = \frac{(x+3)(x+5)}{(x+3)(x+7)} = \frac{x+5}{x+7}. \quad (\S\S\ 115,\ 116)$$

WRITTEN EXERCISE

Reduce to lowest terms:

1. $\frac{x^2+18x+77}{x^2+20x+99}$.
2. $\frac{y^2+16y+48}{y^2+19y+84}$.
3. $\frac{p^2+15p+26}{p^2+18p+65}$.
4. $\frac{m^2+18m+17}{m^2+19m+34}$.
5. $\frac{x^2+16x-80}{x^2+14x-120}$.
6. $\frac{p^2q^2+4pq-77}{p^2q^2+2pq-99}$.
7. $\frac{x^2y^2+8xy-48}{x^2y^2+5xy-84}$.
8. $\frac{x^2-18xy-63y^2}{x^2-17xy-84y^2}$.
9. $\frac{m^2n^2-26mn+105}{m^2n^2-27mn+110}$.
10. $\frac{a^2b^2c^2-15abc+26}{a^2b^2c^2-18abc+65}$.
11. $\frac{p^2q^2-18pqr+17r^2}{p^2q^2-19pqr+34r^2}$.
12. $\frac{m^2n^2+20mn+91}{m^2n^2+21mn+104}$.

WRITTEN EXERCISE

Reduce to fractional forms:

1. $x^3 + x^2 + x + \dfrac{1}{x+1}$.
2. $x^4 - x^2 + 1 - \dfrac{x}{1-x}$.
3. $m + n - p - \dfrac{p}{m-n}$.
4. $m^2n - 13 + \dfrac{mn^2}{mn+2}$.
5. $8\,p^2q^2 - pq + \dfrac{p^3q^3}{pq-1}$.
6. $3\,x^2y - 4\,xy^2 + \dfrac{x^3y^3}{x-y}$.
7. $11x^7y^6 - 17x^5y^3 - \dfrac{17-x^3}{9\,xy}$.
8. $23m^{11} - 17\,n^{12} - \dfrac{m^{15}-n^{12}}{23\,m^4+17}$.

Reduce to entire or mixed quantities:

9. $\dfrac{x^{17}+x^{14}-3}{x^{11}-1}$.
10. $\dfrac{y^7+3\,y^4+2\,y^3}{y^3\,(y^3+1)}$.
11. $\dfrac{x^2+22xy+121}{x+11}$.
12. $\dfrac{x^{12}y^9 - x^9y^6+2}{x^9y^6+1}$.
13. $\dfrac{x^4+4\,x^3y+4\,x^2y^2}{x+y}$.
14. $\dfrac{p^3q^2r^3 - 17\,pqr+3}{pqr-1}$.
15. $\dfrac{m^2n^2 - 17\,mn+15}{mn-5}$.
16. $\dfrac{x^4 - 34\,x^2y^2 + 289\,y^4}{x^2-17\,y^2}$.

Reduce to fractions having the lowest common denominator:

17. $\dfrac{y}{x^2}, \dfrac{y}{x^2+xy}$.
18. $\dfrac{a^2-1}{a^2+a}, \dfrac{a-1}{a^2}$.
19. $\dfrac{x+y}{xy-y^2}, \dfrac{x-y}{xy}$.
20. $\dfrac{x+3\,y}{3x+6\,y}, \dfrac{4x-y}{6xy}$.
21. $\dfrac{ab+1}{a^2b^2-ab}, \dfrac{ab-1}{a^3b^3}$.
22. $\dfrac{p+2q+r}{2\,(p-q)}, \dfrac{p-2q+r}{4}$.
23. $\dfrac{p^2-4pq+q^2}{4\,pq^2+q^3}, \dfrac{p^2+q^2}{2\,q^3}$.
24. $\dfrac{a^2+2\,ab+b^2}{ax-bx}, \dfrac{a^2-2\,ab+b^2}{ax}$.

ORAL EXERCISE

1. Add $\frac{1}{2}$ and $\frac{1}{4}$; $\frac{3}{8}$ and $\frac{1}{2}$; $\frac{2}{a}$ and $\frac{4}{a}$.

2. Add $\frac{a}{2}$ and $\frac{a}{3}$; $\frac{a}{b}$ and $\frac{c}{b}$; $\frac{a}{b}$ and 1.

3. Add $\frac{a}{a+b}$ and $\frac{b}{a+b}$; $\frac{a+b}{c+d}$ and $\frac{a-b}{c+d}$.

4. In adding fractions, what must be the nature of the denominators?

128. Addition of fractions. — Required to add the fractions $\frac{a}{b}$ and $\frac{b}{b+c}$.

1. The least common denominator is evidently $b(b+c)$.

2. $$\frac{a}{b} = \frac{(b+c)a}{(b+c)b} = \frac{ab+ac}{b^2+bc},$$

and $$\frac{b}{b+c} = \frac{b^2}{(b+c)b} = \frac{b^2}{b^2+bc}.$$

3. Therefore the sum $$= \frac{ab+ac+b^2}{b^2+bc}.$$

WRITTEN EXERCISE

1. $\frac{1}{x} + \frac{1}{x+y}$.

2. $\frac{a+b}{a-b} + \frac{b}{a}$.

3. $\frac{m}{m^2+mn} + \frac{1}{m}$.

4. $\frac{x}{x^2-1} + \frac{1}{x}$.

5. $\frac{1}{a} + \frac{1}{b} + \frac{a-b}{ab}$.

6. $\frac{m}{n} + \frac{1}{m+n}$.

7. $\frac{a}{b(a+b)} + \frac{b}{a+b}$.

8. $\frac{p^2+pq}{p(p+q)^2} + \frac{p-q}{p}$.

9. Solve the equation $3x + 17 = x + 59$.

10. A man bought some goods at 6% discount from the marked price, paying $1128. What was the marked price?

11. $\frac{3q}{p^2 - pq} + \frac{3q}{p^2}$.

12. $\frac{bc}{ab + ac} + \frac{1}{abc}$.

13. $\frac{x}{m^2} + \frac{x^2}{m^3 + 4m^2}$.

14. $\frac{a}{b} + \frac{b}{c} + \frac{c}{d} + \frac{d}{e}$.

15. $\frac{x - y}{x^2} + \frac{x^3}{x^2 + xy}$.

16. $\frac{x + y}{x^2} + \frac{x^2 - y}{x + 1}$.

17. $\frac{m + n}{m^2} + \frac{m - n}{m^2 + 2mn}$.

18. $\frac{y}{x^4} + \frac{y}{x^2(x - y)^2}$.

19. $\frac{3p - q}{9p^2 + 6pq} + \frac{p + 3q}{3p}$.

20. $\frac{3}{x^2(x + 2)^2} + \frac{2}{x^2} + \frac{5}{x}$.

Solve the equations in Exs. 21–32:

21. $\frac{x}{2} + \frac{x}{3} = \frac{x}{6} + 120$.

22. $\frac{x}{3} + \frac{x}{5} = \frac{x}{30} + 141$.

23. $18 - \frac{x}{3} = 27 - \frac{x}{0.3}$.

24. $\frac{x + 1}{5} + \frac{x - 4}{5} = \frac{x}{3}$.

25. $4x + 7 = 2x + 41$.

26. $76 - 2x = 93 - 5x$.

27. $32 - x = 48 - 13x$.

28. $41x - 76 = 17x + 20$.

29. $182x - 175 = 157x$.

30. $19.43x - 83.6 = 2.71x$

31. $\frac{x + 2}{3} + \frac{x - 2}{5} = x - 2$.

32. $26 - \frac{2x}{5} = 49 - 6x + x$.

33. Find the number whose half, third, fourth, and sixth together equal 15.

34. What number diminished by 3 equals the sum of the half and fifth of the number?

35. Find the number whose half, fourth, eighth, and sixteenth together equal the number less 2.

36. A man's salary when increased 8% amounted to $1350 a year. What was it before the increase?

129. Subtraction of fractions. — Required, from $\frac{a}{b}$ to subtract $\frac{a-b}{a+b}$.

1. The least common denominator is evidently $b(a+b)$.

2. $$\frac{a}{b} = \frac{(a+b)a}{(a+b)b} = \frac{a^2+ab}{ab+b^2}.$$

3. $$\frac{a-b}{a+b} = \frac{b(a-b)}{b(a+b)} = \frac{ab-b^2}{ab+b^2}.$$

4. The difference $= \dfrac{a^2+ab-(ab-b^2)}{ab+b^2}$

$$= \frac{a^2+ab-ab+b^2}{ab+b^2} = \frac{a^2+b^2}{ab+b^2}.$$

WRITTEN EXERCISE

1. $\dfrac{1}{x} - \dfrac{1}{x+y}$.

2. $\dfrac{1}{x-1} - \dfrac{1}{x}$.

3. $\dfrac{p}{q} - \dfrac{p-q}{p-1}$.

4. $\dfrac{ax}{by} - \dfrac{a-x}{b-y}$.

5. $\dfrac{1}{p+q} - \dfrac{3}{pq}$.

6. $\dfrac{a+b^2}{b} - \dfrac{a}{a^2-b}$.

7. $\dfrac{a}{bc} + \dfrac{b}{ca} - \dfrac{1}{abc}$.

8. $\dfrac{a-b}{(a+b)^2} - \dfrac{a+b}{(a+b)^2}$.

9. $\dfrac{p+q+r}{p-q} - \dfrac{p+q}{p}$.

10. $\dfrac{a^2+a-1}{a+1} - \dfrac{a-1}{a}$.

11. Solve the equation $\dfrac{x}{2} + \dfrac{x}{4} = \dfrac{3}{4}$.

12. Solve the equation $2x + 8\frac{3}{4} = 16\frac{3}{4}$.

13. Solve the equation $\frac{3}{4}x - \frac{1}{16}x = 7007$.

14. Solve the equation $5x + \dfrac{a}{b} = \dfrac{1}{b} + 10 - \dfrac{1-a}{b}$.

15. A man borrowed a sum of money for 1 yr. at 5%. The sum of principal and interest was \$262.50. How much did he borrow?

16. $\frac{x+y}{y+z}-\frac{x}{y^2}$.

17. $\frac{a}{a^2-1}-\frac{1}{a}$.

18. $\frac{p}{pq-q^2}-\frac{p^2}{q^2}$.

19. $\frac{q}{p^2+7p}-\frac{3}{p^2}$.

20. $\frac{x}{x^3+x^2y}-\frac{y}{x^4}$.

21. $\frac{x}{x^2y-y^2}-\frac{2}{y^2}$.

22. $\frac{x^2+x}{4x-6}-\frac{3}{2x}$.

23. $\frac{a}{xy+zx}-\frac{b}{x^2y^2}$.

24. $\frac{a+4}{a^2+3a}-\frac{a-2}{a^2}$.

25. $\frac{x^3}{x^4y-xy^4}-\frac{1}{x^2y^2}$.

26. $\frac{2ab+b^2}{a^2-2ab}-\frac{a+b}{a}$.

27. $\frac{a^2}{6m^2-12m}-\frac{b^2}{3m^2}$.

Solve the equations in Exs. 28–35:

28. $\frac{x}{2}-\frac{x}{3}=5$.

29. $\frac{x}{3}-\frac{x}{5}=4$.

30. $\frac{x}{2}-\frac{x}{7}=x-9$.

31. $\frac{x}{3}-\frac{x}{7}=x-17$.

32. $\frac{x}{5}-2=\frac{x}{7}+2$.

33. $\frac{x}{3}-4=\frac{x}{11}+4$.

34. $\frac{5x}{6}-30=\frac{x}{7}-1$.

35. $\frac{3x}{8}-1=\frac{x}{7}+12$.

36. What number decreased by 14% of itself equals 172?

37. What number increased by 8% of itself equals 37.8?

38. What number increased by 10% of itself equals 687.50?

39. If from a fourth of a certain number I take a seventh of the number, the result is 3. What is the number?

ORAL EXERCISE

1. How much is $\frac{1}{2}$ of $\frac{1}{3}$? $\frac{1}{a}$ of $\frac{1}{b}$? $\frac{2}{3}$ of $\frac{4}{5}$? $\frac{a}{b}$ of $\frac{c}{d}$?

2. How much is $\frac{1}{a}$ of $\frac{1}{a+b}$? $\frac{a}{3}$ of $\frac{x+y}{2}$? $\frac{1}{a+b}$ of $\frac{1}{a-b}$?

3. How much is $3\times\frac{4}{5}$? $a\cdot\frac{b}{c}$? $(a+b)\cdot\frac{a-b}{x+y}$? $\frac{a+b}{x+y}\cdot\frac{a-b}{x-y}$?

4. How do you multiply a fraction by an integer? a fraction by a fraction?

5. In multiplying fractions why should you first *indicate* the multiplication, then factor and cancel, and finally actually multiply?

130. Multiplication of fractions. — Consider the last question (Ex. 5) with the product of $\frac{a^2-ab}{x^2+xy}$ and $\frac{x^2}{a^2}$.

$$\frac{a^2-ab}{x^2+xy}\cdot\frac{x^2}{a^2}=\frac{a^2x^2-abx^2}{a^2x^2+a^2xy},\text{ to be reduced.}$$

If we first indicate the multiplication, and then factor as far as possible and cancel, we have

$$\frac{a(a-b)x^2}{x(x+y)a^2}=\frac{(a-b)x}{(x+y)a},\text{ a reduction easily performed.}$$

WRITTEN EXERCISE

1. $\frac{a}{b}\cdot\frac{bx+by}{ax-ay}$.

2. $\frac{3p^2q}{5xyz}\cdot\frac{xy(x+y)}{9p}$.

3. $(a^2-b^2)\cdot\frac{a+b}{a-b}$.

4. $\frac{2m^2}{n^3}\cdot\frac{n^2x-n^2y}{4m^3+6m^2}$.

5. $\frac{a+5x}{2a}\cdot\frac{10a-15}{5a+25x}$.

6. $(9m^2n^2-16)\cdot\frac{3mn-4}{3mn+4}$.

ORAL EXERCISE

1. Divide $\frac{1}{3}$ by 2; $\frac{1}{a}$ by b; $\frac{1}{x+y}$ by $x-y$.

2. Divide 1 by $\frac{1}{3}$; 2 by $\frac{1}{3}$; a by $\frac{1}{b}$; $m+n$ by $\frac{1}{m-n}$.

3. Divide 5 by $\frac{1}{3}$; 5 by 2; 5 by $\frac{2}{3}$; x by $\frac{a}{b}$; x by $\frac{a}{b+c}$.

4. Divide 5 by $\frac{2}{3}$; $\frac{5}{7}$ by $\frac{2}{3}$; $\frac{a}{b}$ by $\frac{c}{d}$. How do you divide one fraction by another?

131. Division of fractions. — We have already learned that there is a short method of dividing one fraction by another. Just as we divide 3 ft. by 2 in. by reducing to the same denominator (3 ft. $\div$ 2 in. $=$ 36 in. $\div$ 2 in. $= 18$), so we may divide $\frac{3}{4}$ by $\frac{2}{3}$ by reducing to the same denomination ($\frac{3}{4} \div \frac{2}{3} = \frac{9}{12} \div \frac{8}{12} = 9 \div 8 = \frac{9}{8}$). But it is easier to see that $\frac{3}{4} \div \frac{2}{3} = \frac{3}{4} \cdot \frac{3}{2} = \frac{9}{8}$, and that $\frac{a}{b} \div \frac{c}{d} = \frac{a}{b} \cdot \frac{d}{c} = \frac{ad}{bc}$, as already explained on page 72.

Thus, to divide $\frac{4a^3 - 8a^2b}{6x^3 - 9x^2y}$ by $\frac{16a^4b^2}{15x^3y}$, we have, by factoring and multiplying by the reciprocal of the divisor,

$$\frac{15x^3y \cdot 4a^2(a-2b)}{16a^4b^2 \cdot 3x^2(2x-3y)} = \frac{5xy(a-2b)}{4a^2b^2(2x-3y)}.$$

WRITTEN EXERCISE

1. $\frac{a}{b} \div \frac{a+b}{a-b}$.
2. $\frac{x^2+1}{x^2-x} \div \frac{1}{x}$.
3. $\frac{2p}{21q} \div \frac{4p^2}{7q^2}$.
4. $\frac{1}{x-1} \div \frac{y}{x}$.
5. $\frac{a}{3b} \div \frac{a+b}{b}$.
6. $\frac{a^7b}{a^7-b} \div \frac{ab^7}{b-a^7}$.

EQUATIONS INVOLVING FRACTIONS

ORAL EXERCISE

1. By what numbers could you multiply $\frac{1}{3}$ and have the product an integer? also $\frac{2}{3}$; $\frac{5}{7}$; $\frac{a}{b}$; $\frac{m}{m+n}$?

2. What is the smallest number by which we can multiply $\frac{x}{2}$ to have the product an integer? also $\frac{x}{a}$; $\frac{x}{b+c}$?

3. If $\frac{x}{2} = 7$, what is the value of x? By what number do we multiply both members of the equation?

4. If $\frac{x}{a} = b$, what is the value of x? Also find the value of x if $\frac{x}{a+b} = a - b$; if $\frac{x}{m+n} = m + n$.

132. Illustrative problem. — Solve the equation $\frac{x}{5} + 7 = 9$.

1. Since $\frac{x}{5} + 7 = 9$,
2. Therefore $\frac{x}{5} = 2$, by Axiom 3. (State it.)
3. Therefore $x = 10$, by Axiom 4. (State it.)

WRITTEN EXERCISE

Solve the equations in Exs. 1–9:

1. $\frac{x}{7} + 3 = 7$. **2.** $\frac{x}{10} - 7 = 9$. **3.** $\frac{3x}{11} - 6 = 3$.

4. $\frac{2}{7}x + 7 = 15$. **5.** $\frac{2}{3}x + 6 = 42$. **6.** $\frac{3}{7}x - 7 = 83$.

7. $\frac{3}{5}x + 7 = \frac{1}{5}x + 23$. **8.** $\frac{5}{8}x - 6 = \frac{3}{8}x - 5$.

9. If from $\frac{3}{4}$ of the number in our class I take 9, the result is half of the class. How many have we?

ORAL EXERCISE

1. If $\frac{1}{3}x = 7$, what does x equal?
2. If $\frac{2}{3}x = 8$, what does $2x$ equal? x?
3. If $\frac{3}{5}x = 2$, what does $3x$ equal? x?

By what should we multiply both members in order that there shall be no fractions in the following equations?

4. $\frac{5x}{7} + \frac{2x}{3} = 6.$
5. $\frac{2x}{3} + \frac{4x}{5} = \frac{2}{15}.$
6. $\frac{x}{2} + \frac{x}{3} + \frac{1}{4} = 7.$
7. $\frac{x}{3} + \frac{x}{6} + \frac{x}{12} = \frac{3}{4}.$
8. In general, by what should we multiply both members of any equation in order that there shall be no fractions in the resulting equation?

133. Clearing of fractions. — Multiplying both members of an equation by such a number as to have no fractions in the result is called *clearing an equation of fractions.*

To clear an equation of fractions, multiply both members by the least common multiple of the denominators.

134. Illustrative problem. — Solve the equation

$$3 + \frac{2}{3}x = \frac{4}{5}x + 2.$$

1. Clearing of fractions by Axiom 4, multiplying by 15,

$$45 + 10x = 12x + 30.$$

2. Subtracting both 45 and $12x$, so as to place all the x's on one side and all the known terms on the other,

$$10x - 12x = 30 - 45,$$

or

$$-2x = -15.$$

3. Dividing by -2, to find the value of x,

$$x = \frac{15}{2}, \text{ or } 7\tfrac{1}{2}.$$

Check. $3 + \frac{2}{3} \cdot \frac{15}{2} = \frac{4}{5} \cdot \frac{15}{2} + 2$; for each equals 8.

WRITTEN EXERCISE

Solve the equations in Exs. 1–10:

1. $\frac{x}{3}+\frac{x}{2}=7.$

2. $\frac{x}{5}+\frac{x}{3}=20.$

3. $\frac{2x}{5}+3=\frac{x}{3}+4.$

4. $x+\frac{x}{3}=\frac{1}{3}+7.$

5. $\frac{2x}{9}+\frac{x}{6}=\frac{x}{18}+\frac{1}{3}.$

6. $\frac{3x}{7}-\frac{1}{3}=\frac{x}{21}+5.$

7. $4x+\frac{1}{7}=\frac{1}{7}x+4.$

8. $3x+4x=\frac{3}{4}x+\frac{1}{8}$

9. $10+0.1x=5+\frac{1}{8}x.$

10. $0.1x+6.2=0.3x+0.2.$

11. Find a number whose half, third, and fourth added together equals 36.

12. A man spends every year \$200 more than half his salary. In 3 years he saves \$900. How much is his salary?

13. After selling $\frac{1}{2}$ of his farm, and then $\frac{1}{8}$ of what was left, a man still had 140 acres. How many acres had he at first?

14. There is a certain number such that its fourth added to its fifth equals one less than its half. What is the number?

15. A man invests half of a certain sum at 6% interest, and the other half at 5%. The total interest for a year is \$66. How much did he invest?

Solve the equations in Exs. 16–21:

16. $\frac{x}{4}-\frac{x}{16}=\frac{x}{8}+2.$

17. $\frac{10}{x}+\frac{15}{3x}=3.$

18. $3-\frac{x}{2}=\frac{3x}{4}-97.$

19. $\frac{8}{x}+\frac{16}{x}+\frac{4}{x}=14.$

20. $\frac{2x}{3}-\frac{x}{17}=x-20.$

21. $\frac{x}{6}+\frac{x}{12}-\frac{x}{15}=11.$

135. Illustrative problems. — 1. Solve the equation

$$\frac{x+3}{x-5} = 1.8.$$

1. Clearing of fractions by multiplying both members by $x - 5$,

$$x + 3 = 1.8\,x - 9.$$

2. Subtracting $1.8\,x$ and 3 from both members, or transposing,

$$-0.8\,x = -12.$$

3. Dividing both numbers by -0.8,

$$x = 15.$$

2. Two numbers have the ratio of 2 to 3. If 3 be subtracted from the larger, the result is 7 more than the smaller. What are the numbers?

1. Because $\frac{2\,x}{3\,x} = \frac{2}{3}$, we may conveniently take $2\,x$ and $3\,x$ to represent the numbers.

2. Then $3\,x - 3 =$ the larger less 3,
and $2\,x + 7 =$ the smaller plus 7.

3. Therefore $3\,x - 3 = 2\,x + 7.$

4. Therefore $3\,x - 2\,x = 7 + 3$, by Axioms 2 and 3 (transposing), or $x = 10.$

5. But we wish $2\,x$ and $3\,x$, and these are 20 and 30.

Check. $30 - 3 = 27 = 20 + 7.$

WRITTEN EXERCISE

Solve the equations in Exs. 1–6:

1. $\frac{x}{x+4} = \frac{4}{5}.$
2. $\frac{x-1}{x+1} = \frac{7}{8}.$
3. $\frac{x+7}{x-7} = 2.$
4. $\frac{120}{x-4} = 30.$
5. $\frac{9-x}{3\,x} = 1\frac{1}{6}.$
6. $\frac{x-2}{4} = \frac{x+2}{8}.$

136. General directions. — We have now found the general plan of solving a linear equation:

1. *Clear of fractions.*
2. *Transpose the* x*'s to the left side and the known terms to the right.*
3. *Collect the terms and divide by the coefficient of* x.

For example, solve the following:

1. $$\frac{x}{a-b}+\frac{2x}{a}+7=8.$$

It will evidently save some work to transpose the 7 first.

2. Then $$\frac{x}{a-b}+\frac{2x}{a}=1.$$

3. Clearing of fractions by multiplying by $a(a-b)$,
$$ax+2ax-2bx=a(a-b).$$

4. Simplifying and collecting terms,
$$(3a-2b)x=a(a-b).$$

5. Dividing by the coefficient of x,
$$x=\frac{a(a-b)}{3a-2b}, \text{ or } \frac{a^2-ab}{3a-2b}.$$

WRITTEN EXERCISE

Solve the equations in Exs. 1–8:

1. $\frac{1}{2}x+7=\frac{3}{8}x+18.$

2. $\frac{3}{5}x-\frac{2}{3}x=\frac{4}{7}-\frac{2}{3}.$

3 $\frac{x}{a+1}+\frac{x}{b}=1.$

4. $\frac{a+b}{x}+\frac{a-b}{x}=2a.$

5. $\frac{x}{2a}+\frac{x+1}{a+b}=3.$

6. $\frac{x}{4a}+\frac{3x}{5a}+\frac{4x}{a}=97.$

7. $\frac{x+3}{x-1}+1+a=0.$

8. $\frac{2x+1}{a}+3=\frac{3x-1}{a+b}.$

9. A flag pole is so broken off by the wind that $\frac{1}{3}$ of the part broken off equals $\frac{1}{4}$ of the part left standing. The original height was 90 ft. How much was broken off?

137. Illustrative problem. — Rob can gather the apples on his father's trees in 6 hours, and Tom can do it in 5 hours. Working at the same rates, how long will it take the two together to gather them?

1. Rob can gather $\frac{1}{6}$ of them in 1 hour, and
   Tom can gather $\frac{1}{5}$ of them in 1 hour.

2. If it takes the two together x hours, they can gather $\frac{1}{x}$ of them in 1 hour.

3. Therefore $\quad \frac{1}{6}+\frac{1}{5}=\frac{1}{x}.$

4. Therefore $5x+6x=30$, by clearing of fractions,

or $\quad 11x=30.$

$$x=2\tfrac{8}{11}.$$

Therefore it takes the two together $2\frac{8}{11}$ hours.

Solve the equations in Exs. 1–6:

1. $\frac{1}{7}+\frac{1}{8}+\frac{1}{4}=\frac{1}{x}.$ 2. $\frac{x}{2}+\frac{3x}{4}+\frac{5x}{8}=30.$

3. $\frac{x}{a}+\frac{cx}{d}+e=b.$ 4. $\frac{x+a}{b}-\frac{x-b}{a}=\frac{x+c}{ab}.$

5. $\frac{x-1}{2}-\frac{1-x}{3}-\frac{x-2}{6}=\frac{29}{6}.$ 6. $\frac{x-1}{7}+\frac{x+1}{8}+\frac{x}{3}=9.$

7. The sum of two numbers is 90, and one is 20% less than the other. What are the numbers?

8. A piece of cloth lost 20% of its length in shrinking, and was then 60 yd. long. How long was it originally?

9. The perimeter of a certain rectangle is 200 inches, and the length is half as much again as the width. Required the dimensions.

10. A man sold a pony for $\frac{1}{4}$ more than it cost, and the buyer sold it for $\frac{1}{5}$ less than he paid for it, receiving $60. How much did it cost the first man?

11. The sum of two numbers is 50, and the less is $\frac{2}{3}$ the greater. What are the numbers?

12. If 110 be divided by 1 less than a certain number, the result is $36\frac{2}{3}$. What is the number?

13. If 175 be divided by 3 less than five times a certain number, the result is $3\frac{34}{47}$. What is the number?

14. If to $\frac{1}{35}$ of the sum of 3 and three times a certain number there be added $\frac{1}{12}$ of the number, the result is 5. What is the number?

15. A man spent $150 more than half of his income each year for two years. He saved $800 in the two years. What was his annual income?

16. The distance around a certain rectangular field is 78 rd., and the length is 2.9 times the width. What is the width? the length? the area?

17. The profit to be divided among three partners is $10,000. B receives $2\frac{1}{3}$ times as much as A, and A receives 30% as much as C. How much does each receive?

18. The perimeter of a triangle is 8.1 in.; the shortest side is $\frac{1}{2}$ the longest one, and the longest one is $33\frac{1}{3}$% longer than the third side. What is the length of each side?

19. Albany is $\frac{1}{3}$ of the way from New York to Buffalo, Rochester is $\frac{4}{5}$ of the way from Albany to Buffalo, and it is 60 miles from Rochester to Buffalo. Required the distance from New York to Buffalo.

20. A man started in business with a certain capital. He gained $1000 the first year, lost half of what he then had the second year, and gained $2000 the third year. He then found that he had the same amount with which he started. How much was it?

PROPORTION

138. Ratio. — The relation of one quantity to another of the same kind, as expressed by division, is called the *ratio* of the first to the second.

139. Proportion. — An expression of the equality of two ratios is called a *proportion.*

The ratio of 2 in. to 5 in. may be indicated thus 2 in. : 5 in., or $\frac{2 \text{ in.}}{5 \text{ in.}}$. The equality of this ratio to the ratio of 4 ct. to 10 ct. may be indicated thus: 2 in. : 5 in. = 4 ct : 10 ct., or $\frac{2 \text{ in.}}{5 \text{ in.}} = \frac{4 \text{ ct.}}{10 \text{ ct.}}$.

140. A proportion is an equation. — The proportion $x : 2 = 14 : 4$ may be written $\frac{x}{2} = \frac{14}{4}$, a simple or linear equation. Therefore *a proportion is merely an equation containing fractions, and is best solved like an equation.*

141. Extremes and means. — In a proportion, the first and last terms are called the *extremes*, and the second and third the *means.*

Thus, in the proportion $x . a = b \cdot c$, x and c are the extremes and a and b the means.

WRITTEN EXERCISE

1. Write the proportion $x . a = b : c$ in fractional form.

2. Using this fractional form, clear of fractions and show that $xc = ab$.

3. In the same way, using the fractional form, show that $x = \frac{ab}{c}$.

4. From Ex. 2, write out a statement concerning the product of the extremes equaling some other product.

142. Relation of extremes to means. — On the preceding page it was proved in Exs. 2 and 3 that, in any proportion,

1. *The product of the extremes equals the product of the means.*

2. *The product of the means divided by one extreme equals the other extreme.*

143. Terms may be considered abstract. — In the proportion 2 in. : 3 in. = 5 ct. : $7\frac{1}{2}$ ct., it is of course impossible to multiply inches and cents together. But because

$$\frac{2\text{ in.}}{3\text{ in.}} = \frac{2}{3}, \text{ and } \frac{5\text{ ct.}}{7\frac{1}{2}\text{ ct.}} = \frac{5}{7\frac{1}{2}}, \text{ we see that}$$

The terms of a proportion may all be considered as abstract.

144. Formerly, before the equation was common in the school, proportions were solved by rule 2, above.

145. Illustrative problems. — 1. Solve the proportion

$$x : 9 = 34 : 153.$$

1. Writing this $\frac{x}{9} = \frac{34}{153}$, we have a simple equation.

2. Therefore $x = \frac{9 \cdot 34}{153} = 2.$ (with 9 and 153 cancelled, leaving 17)

2. Solve the proportion $12 : x = 27 : 18$.

1. That is, $\frac{12}{x} = \frac{27}{18} = \frac{3}{2}.$

2. Clearing of fractions, $24 = 3x$, and $8 = x$.

WRITTEN EXERCISE

Solve the following:

1. $x \cdot 84 = 9 : 63.$
2. $37 : x = 259 : 77.$
3. $51 : 21 = x : 7.$
4. $245 : 75 = 98 : x.$
5. $19 : 33 = x : 231.$
6. $x : 125 = 25 : 625.$

146. Lever and fulcrum. — If we take a yardstick and balance it, as shown in the picture, with weights at X and Y, we have a *lever*. The point F is called the *fulcrum*.

147. Law of the lever. — Now if we put F 25 in. from X, it will be 11 in. from Y, because the stick is 36 in. long. We then find that the weight x at X will have to the weight y at Y the ratio $11:25$; that is, $\frac{x}{y}=\frac{11}{25}$, or

The weight at X *is to the weight at* Y *as the distance of* F *from* Y *is to its distance from* X.

This law, easily proved in class, is called the *law of the lever*.

For example, if a 6-ft. lever has a fulcrum 1 ft. from the weight W, and if p is the power we must use to lift w, we know that $\frac{w}{p}=\frac{5}{1}$; that is, $p=\frac{1}{5}w$. Hence to lift 100 lb. we need exert a power of only 20 lb.

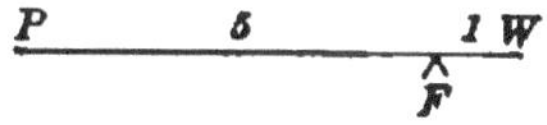

WRITTEN EXERCISE

1. Where must we place the fulcrum under a 12-ft. plank that a 56-lb. boy may balance one who weighs 112 lb.?

2. A father puts the fulcrum 2 ft. from his end of a 10-ft. plank and just balances his son, who weighs 40 lb. How much does the father weigh?

3. Two boys balance a seesaw, the plank being 12 ft. long. The fulcrum is 5 ft. from the heavier boy, who weighs 105 lb. How much does the other weigh?

4. If Rob has an iron bar $4\frac{3}{4}$ ft. long, and wishes to pry up a 300-lb. rock, and he weighs 80 lb., how far from the stone must he place the fulcrum, making no allowance for taking hold of the bar or reaching under the stone?

148. Illustrative problem. — If 6 sheep cost \$30, how much do 8 sheep cost at the same rate?

1. Because the ratio of the cost equals the ratio of the number, then, if $x =$ the number of dollars paid for 8 sheep,

2. $$x : \$30 = 8 \text{ sheep} : 6 \text{ sheep},$$

or $$x : 30 = 8 : 6,$$

or $$\frac{x}{30} = \frac{8}{6}.$$

3. Therefore $$x = \frac{30 \cdot 8}{6} = 5 \cdot 8 = 40.$$

4. Therefore the 8 sheep cost \$40.

Teachers will find it much better to encourage pupils to put the unknown term first, as in other equations, instead of last

WRITTEN EXERCISE

1. If 9 yd. of carpet cost \$14.40, how much will 7 yd. of the same carpet cost?

2. If I walk 2 mi. in 50 min., how long will it take me to walk $2\frac{3}{4}$ mi. at the same rate?

3. If a train travels 105 mi. in $2\frac{1}{2}$ hr., how far will it travel in $1\frac{1}{2}$ hr. at the same rate?

4. A wheel 2 ft. in diameter has a circumference of $6\frac{2}{7}$ ft. What is the circumference of a wheel 5 ft. in diameter?

5. A wheel having a circumference of 10 ft. has a radius of $1\frac{13}{22}$ ft. What is the radius of a wheel 7 ft. in circumference?

6. A man timed the express train that he was on and found that it made 2 mi. in 2 min. 8 sec. It was then going at what rate per hour?

7. If the work on a tunnel progresses at the rate of 16 ft. a week, and the tunnel is to be 1520 ft. long, how many weeks will it take to complete it?

ORAL EXERCISE

1. If it takes you 4 min. to clean a blackboard, how long will it take you to clean half of it?

2. If it takes you 4 min. to do the work, how long will it take two of the class at the same rate?

3. If it takes 2 men 14 days to excavate a cellar, how long will it take twice as many men?

149. Varying inversely. — In these examples, as we double the number of men we divide the time by 2. That is, the time decreases in the same ratio that the workmen increase, or the time *varies inversely* as the men.

150. Illustrative problem. — If it takes 20 days for 15 men to do a piece of work, how long will it take 10 men to do it?

1. The ratio of the times is $\frac{x}{20}$.

2. Because 10 men do it in x days and 15 men in 20 days, therefore the ratio of the men is $\frac{10}{15}$, and the *inverse ratio* is $\frac{15}{10}$.

3. Therefore $\frac{x}{20} = \frac{15}{10}$.

4. Therefore $x = \frac{20 \cdot 15}{10} = 30$, and it takes 10 men 30 days.

WRITTEN EXERCISE

1. I have enough grain to last my 2 horses 3 months. If I buy another horse, how long will the grain last?

2. With a certain quantity of wool I can make 42 yd. of goods 24 in. wide. How many yards can I make if it is 1 yd. wide?

3. It takes 56 yd. of 27-in. carpet for my parlor. How much will be needed if I use carpet 1 yd. wide, assuming that it will cut as economically?

WRITTEN EXERCISE

1. If 17 horses cost $1870, how much will 11 horses cost at the same rate?

2. If 35 men can do a job of paving in 8 days, how long will it take 56 men at the same rate?

3. How long would it take 50 head of cattle to eat the same amount of fodder that 30 head eat in 45 days?

4. If the interest on a certain sum is $112.50 for 1 yr. 6 mo., how much is it for 2 yr. 3 mo. at the same rate?

5. An automobile is timed, and found to make 450 yd. in 30 sec. At this rate, how long will it take it to go a mile?

6. If 150 mi. of railroad cost $3,600,000, how much more will it cost to extend the road 36 mi. farther at the same rate?

7. It costs me 75 ct. a night to light my store by electricity when the lights burn 5 hours. How much will it cost when they burn 7 hours?

8. It is estimated that it will cost $185,000 to put a state road through a hilly part of the country for a distance of 37 miles. How much will it cost for 150 miles at the same rate?

9. There were two pieces of sodding, of the same size, to be done in our park. It took 3 men 5 days to do the first piece. How long will it take the 2 men whom we now have to do the second?

10. A builder had 4 plasterers at work 12 days in plastering one of the stories of an apartment house. How long will it take 6 men to plaster the next story? If the builder puts 8 men at work on the next story, how long will it take them? The rooms are the same for all stories.

151. Similar figures. — Figures which have exactly the same shape are called *similar figures.*

For example, two circles are similar figures; also two squares, two equilateral triangles, two cubes, or two spheres.

ORAL EXERCISE

1. What is the area of a square 2 in. on a side? also of one that is twice as long? (Draw the pictures if necessary.)

2. The diagonal of a square 1 in. on a side is 1.4 in. How long is the diagonal of a square 2 in. on a side? (If in doubt, measure it.)

3. What is the volume of a cube that is 1 in. on an edge? also of one that is 2 in. on an edge? (Build the latter of inch cubes if necessary.)

4. If one side of an equilateral triangle is 3 in., what is the perimeter of another equilateral triangle that is twice as high? three times as high?

152. Proportion related to similar figures. — We may infer from the above exercise that, in similar figures,

1. *Corresponding lines are proportional.*

That is, if the radius of one circle is twice that of another, the circumference of the one is twice that of the other.

2. *Areas are proportional to the squares of corresponding lines.*

That is, if one equilateral triangle is twice as high as another, the area is 2^2, or 4, times that of the other.

3. *Volumes are proportional to the cubes of corresponding lines.*

That is, if the radius of one sphere is twice that of another, the volume of the one is 2^3, or 8, times that of the other.

153. Illustrative problem. — If the area of a circle is 3.8 sq. in., what is the area of a circle of twice the diameter?

1. Since the areas are proportional to the squares of corresponding lines,

$$x : 3.8 = 2^2 : 1,$$

or

$$\frac{x}{3.8} = \frac{4}{1}.$$

2. Multiplying by 3.8, $x = 15.2$.

3. Therefore the area is 15.2 sq in.

WRITTEN EXERCISE

1. A cylindrical can holds a pint. How much will a similar one hold if it is $1\frac{1}{2}$ times as high?

2. A box 3 in. long has a volume of 10.5 cu. in. What is the volume of a box of the same shape, 4 in. long?

3. A toy balloon 6 in. in diameter has a volume $113\frac{1}{7}$ cu. in. If it is inflated to 7 in. in diameter, what will be the volume?

4. A triangle has its sides 3 in., 4 in., 5 in. Another triangle of the same shape has its shortest side 2 in. What are the lengths of the other sides?

5. Of the two triangles in Ex. 4, the first has an area of 6 sq. in. What is the area of the second?

6. A certain projectile for a man-of-war's gun weighs 250 lb. What is the weight of a similar one of which the length is 10% more?

7. Of the two projectiles mentioned in Ex. 6, the area of the surface of the second is how many times that of the first?

8. A photograph in which a house appears as 1.7 in. high, and a tree as 1.5 in. high, is enlarged so that the house appears as 2.1 in. high. How high does the tree appear?

154. Proportional parts. — The profits of a business are to be divided between two partners in the ratio of 3 : 5. The profits for a year are $4624. What is the share of each?

1. If $x =$ the number of dollars in the smaller share,
 $4624 - x =$ " " " " " " larger "
2. Therefore $x : (4624 - x) = 3 : 5,$

or
$$\frac{x}{4624 - x} = \frac{3}{5}.$$

3. Clearing of fractions, $5x = 3 \cdot 4624 - 3x.$
4. Adding $3x$ to these equals,
 $8x = 3 \cdot 4624.$
5. $x = 3 \cdot 578 = 1734$, the number of dollars in the smaller share.
6. $4624 - 1734 = 2890$, " " " " " " larger "

155. While this furnishes a good exercise in proportion, teachers will naturally encourage pupils to say that out of 8 parts, 3 go to one and 5 to the other. Hence one gets $\frac{3}{8}$ and the other $\frac{5}{8}$ of the $4624 The treatment of partitive proportion as a distinct subject is unnecessary.

WRITTEN EXERCISE

1. Divide $375 in the ratio 2 · 3.

2. Divide $2619 in the ratio 7 : 2.

3. Divide 1638 ft. in the ratio 5 : 13.

4. Divide $4837 between two partners in the ratio 3 : 4.

5. An alloy contains 19 parts copper to 11 parts tin. How many pounds of each in 570 lb. of the alloy?

6. A certain powder contains 2 parts of saltpeter to 1 of charcoal and sulphur. How many pounds of saltpeter in 462 lb. of powder?

7. Air contains 21 parts of oxygen to 79 parts of nitrogen. How many cubic feet of each in a schoolroom 22 ft. long, 18 ft. wide, and 12 ft. high?

WRITTEN EXERCISE

1. If $34 : x = 51 : 27$, find the value of x.

2. If $76 : 133 = x : 77$, find the value of x.

3. If $52 : 91 = 84 : x$, find the value of x.

4. If $\frac{3}{7} : 1 = x : 3\frac{1}{2}$, find the value of x.

5. If $\frac{3}{16} : x = x : 5\frac{1}{3}$, find the value of x.

6. What number has to 5.81 the ratio of 5 to 7?

7. To what number has 15.2 the ratio of 1.9 to 2?

8. Divide $1025 into two parts in the ratio of 11 to 30.

9. If the surface of one sphere is 7.2 sq. in., what is the surface of another sphere of twice the radius?

10. If the area of one equilateral triangle is 8.2 sq. in., what is the area of another one with 4 times the perimeter?

11. If the interest on a certain sum for 1 yr. 2 mo. 10 da. is $21.50, what is the interest on the same sum at the same rate for 6 mo.?

12. The sum of two numbers is 30, and the ratio of one to the other is $2 : 3$. What are the numbers? (Let $2x$ equal one number.)

13. Four partners divide the year's profits, $12,250, so that A's share is 25% of D's, D's is twice B's, and C's is as much as A's and B's together. Find each share.

14. Three partners divide the year's profits, $6000, so that the first receives $1 to the second's $1.50, and the second receives $1 to the third's $2.33⅓. How much does each receive?

15. Five boys go fishing. They catch 80 fish, A and C catching the same number, and B catching ⅔ as many as C. D catches as many as A and B together, and E catches as many as B and D together. How many does each catch?

SQUARE ROOT

156. Square root. — If a number or an algebraic expression has two equal factors, one of these factors is called its *square root.*

For example, because $4 = 2 \cdot 2$, therefore 2 is a square root of 4; and because 4 also equals $-2 \cdot -2$, therefore -2 is also a square root of 4.

Also, because $a^2 + 2ab + b^2 = (a+b)^2$, therefore $a + b$ is a square root of $a^2 + 2ab + b^2$. In the same way $-(a+b)$ is also a square root.

157. Law of signs. — Because $+a \cdot +a = a^2$, and $-a \cdot -a = a^2$, therefore a^2 has two square roots, $+a$ and $-a$. That is,

The square root of a quantity is either positive or negative.

158. Symbols. — The square root of a^2 is indicated by $\sqrt{a^2}$. Hence $\sqrt{4} = 2$, $\sqrt{16} = 4$. But since a square root has two signs, the double sign $\pm$, read "plus or minus" or "positive or negative," is used, thus: $\pm\sqrt{9} = \pm 3$.

159. Illustrative problem. — What is the square root of 144?

1. We see that the factors of 144 are

$$2, 2, 2, 2, 3, 3.$$

2. Separating these into two equal groups, we see that

$$144 = 2 \cdot 2 \cdot 3 \times 2 \cdot 2 \cdot 3 = 12 \times 12.$$

3. Therefore the square roots are $+12$ and -12.

$$\begin{array}{r} 2\,)\underline{144} \\ 2\,)\underline{72} \\ 2\,)\underline{36} \\ 2\,)\underline{18} \\ 3\,)\underline{\;\;9} \\ 3 \end{array}$$

WRITTEN EXERCISE

Find the square roots of the following:

1. 625. 2. 576. 3. 225. 4. 441. 5. 729.
6. 121. 7. 169. 8. 289. 9. 361. 10. 1024.

160. Monomials. — Of course the square root of a monomial square can easily be found.

For example, because $a^4 = a^2 \cdot a^2,$
therefore $\sqrt{a^4} = a^2.$
Likewise $\sqrt{a^4b^6x^2} = a^2b^3x.$

161. Trinomials. — Because $(a+b)^2 = a^2 + 2\,ab + b^2$, therefore we can easily find the square root of any expression of this form.

For example, because $a^2 - 6\,ab + 9\,b^2 = a^2 + 2\,a(-3\,b) + (-3\,b)^2 = (a-3\,b)(a-3\,b)$, therefore $\sqrt{a^2 - 6\,ab + 9\,b^2} = a - 3\,b.$

ORAL EXERCISE

State the square roots of the following:

1. $a^6b^6c^2.$
2. $a^4m^6x^2.$
3. $a^8b^6c^4.$
4. $9\,a^2b^4.$
5. $16\,x^8y^6.$
6. $25\,a^6b^4x^2.$
7. $49\,a^2b^2c^2.$
8. $64\,x^4y^2.$
9. $81\,m^4n^4x^4.$
10. $100\,x^2y^4z^4.$
11. $121\,a^2b^4x^6.$
12. $144\,x^2y^2z^2.$

WRITTEN EXERCISE

State the square roots of the following:

1. $x^2 + 6\,x + 9.$
2. $9 + 6\,x^2 + x^4.$
3. $4\,a^4 + 4\,a^2 + 1.$
4. $m^4 + 4\,m^2 + 4.$
5. $9\,x^4 + 6\,x^2 + 1.$
6. $x^2 - 14\,x + 49.$
7. $a^2 + 12\,a + 36.$
8. $4\,a^2 + 4\,ab + b^2.$
9. $m^2 + 18\,m + 81.$
10. $x^4 + 2\,x^2y^2 + y^4.$
11. $9\,m^2 + 6\,mn + n^2.$
12. $p^2 + 10\,pq + 25\,q^2.$
13. $x^2 - 12\,xy + 36\,y^2.$
14. $x^2 - 16\,xy + 64\,y^2.$
15. What is the side of a square whose area is $a^2 + 2\,ab + b^2$? $a^2 - 2\,ab + b^2$? $x^2 + 2\,x + 1$?

162. The square root of numbers. — Required the square root of 2209.

If we let f = the *f*ound part of the root at any time,
and n = the *n*ext figure of the root,
then $(f+n)^2 = f^2 + 2fn + n^2$.

Therefore if we take away f^2 we shall have $2fn + n^2$, and if we divide this by $2f$ we shall find nearly n.

Since the entire explanation of square root depends on this fact, teachers are advised to see that it is clearly understood, both from the figure and from the formula, before proceeding.

	47	$= f + n$, the root	
	2209	contains $f^2 + 2fn + n^2$	
$f^2 =$	1600		
$2f = 80$	609	"	$2fn + n^2$
$2f + n = 87$	609	=	" "

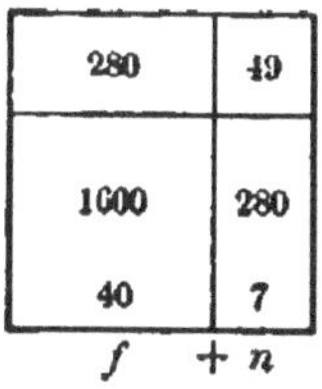

280	49
1600	280
40	7
f	$+n$

The greatest square of 10's in 2209 is 1600. This is f^2, and therefore $f = 40$.

Then 609 contains $2fn + n^2$, because f^2 has been subtracted.

Dividing this by $2f$, we shall have nearly n. Hence $n = 7$.

But $2f + n$ multiplied by n equals $2fn + n^2$. Therefore we have taken $f^2 + 2fn + n^2 = 40^2 + 2 \times 40 \times 7 + 7^2 = 47^2$ from 2209.

Therefore 47 is the square root of 2209.

See also the figure, where it appears that the large square equals the sum of $40^2 + 2 \times 40 \times 7 + 7^2$.

$$\begin{aligned} f^2 &= 1600 \\ 2fn &= \begin{cases} fn = 280 \\ fn = 280 \end{cases} \\ n^2 &= 49 \\ & \overline{2209} \end{aligned}$$

If we separate the number into periods of two figures each, beginning at the decimal point, we shall find the number of integral places in the root, but it is not necessary.

WRITTEN EXERCISE

Find the square root of the numbers in Exs. 1–5:

1. 3249. 2. 3721. 3. 3969. 4. 5041. 5. 6241.

163. Square root with decimals. — Required the square root of 151.29.

The greatest square of 10's in 151.29 is 100. This is f^2, and therefore $f = 10$.

Then 51.29 contains $2fn + n^2$. (Why is this?)

Dividing by $2f = 20$, we find $n = 2$.

We have now found $f + n = 12$, the square being $100 + 44 = 144$.

Since 12 has been found, let us call this f (for *found*). Of course this is not the same as the first number found; it is larger, because we have found more.

	12.3
	151.29
	100.
$2f = 20$	51.29
$2f + n = 22$	44.
$2f = 24$	7.29
$2f + n = 24.3$	7.29

7.29 contains $2fn + n^2$, because we have subtracted $f^2 = 144$.

Dividing by $2f = 24$, we find $n = 0.3$.

$2f + n$ multiplied by n equals $2fn + n^2$, the rest of the square.

WRITTEN EXERCISE

Extract the square roots in Exs. 1–17:

1. 80.4609. 2. 1944.81. 3. 1.1025.

4. 0.117649. 5. 0.822649. 6. 0.2809.

In Exs. 7–9 find first the greatest square of 100's.

7. 12,321. 8. 54,756. 9. 63,001.

In Exs. 10–12 find first the greatest square of 1000's.

10. 21,224,449. 11. 49,112,064. 12. 96,275,344.

In Exs. 13–17 carry the root to two decimal places only.

13. 2. 14. 5. 15. 7. 16. 8. 17. 11.

18. What is the value of $(\frac{1}{2})^2$? $\sqrt{\frac{1}{4}}$? $\sqrt{\frac{4}{9}}$? $\sqrt{\frac{1089}{2500}}$?

19. Write out a rule for finding the square root of a common fraction. Apply it to finding the value of $\sqrt{\frac{144\,a^4}{169\,b^6}}$.

QUADRATIC EQUATIONS

164. Quadratic equation. — An equation containing the second, but no higher, power of the unknown quantity is called a *quadratic equation.*

165. Complete quadratic. — The equation $x^2 + ax + b = 0$ is a *complete quadratic equation.*

166. Incomplete quadratic. — The equation $x^2 + b = 0$, lacking the first power of the unknown quantity, is called an *incomplete quadratic equation.*

The terms *affected quadratic* for the complete and *pure quadratic* for the incomplete equation are still used, but are being discarded by the better writers.

167. Axiom 6. *Like roots of equal quantities are equal.*

168. Illustrative problem. — Solve the equation $x^2 - 39 = 130$.

1. Adding 39, $x^2 = 169$.
2. Extracting the square root, $x = \pm 13$, by axiom 6.

Check. $(\pm 13)^2 - 39 = 169 - 39 = 130$.

WRITTEN EXERCISE

1. Solve the equation $\frac{1}{2} x^2 = 72$.

2. Solve the equation $\frac{3}{4} x^2 - 7 = 140$.

3. What is the side of a square whose area is 225 sq. in.?

4. What is the value of x if x^2 added to 60 equals 256?

5. Solve the equation $5x^2 - 7x + 75 = 300 - 7x + 4x^2$.

6. Find the number which multiplied by the next higher number equals 144 increased by the number.

7. Find the number which multiplied by the next lower number equals 169 diminished by the number.

169. Illustrative problem. — Solve the equation $x(16x-7) = 7(2800-x)$.

1. Performing the multiplications,

$$16x^2 - 7x = 7 \cdot 2800 - 7x.$$

2. Adding $7x$, $$16x^2 = 7 \cdot 2800.$$

3. Dividing by 16, $$x^2 = 7 \cdot 175 = 7 \cdot 7 \cdot 25.$$

4. Extracting the square root,

$$x = \pm 7 \cdot 5 = \pm 35.$$

Check. $35(16 \cdot 35 - 7) = 7(2800 - 35)$,

for $$19355 = 19355.$$

WRITTEN EXERCISE

1. Solve the equation $(x+6)(x-6) = 864$.

2. Solve the equation $\frac{2}{3}(10+x)(x-10) = 1000$.

3. Solve the equation $4x(1+x) = 2496 + 4(x+1)$.

4. In x seconds an object will fall $16x^2$ ft. How long will it take an object to fall 400 ft.?

5. A library is $1\frac{1}{2}$ times as long as it is wide, and its area is 150 sq. ft. What are the dimensions?

6. The width of a schoolroom is 90% of its length. The area is 360 sq. ft. What are the dimensions?

7. The area of a circle equals $3\frac{1}{7}$ times the square of its radius. What is the radius of a circle whose area is 154 sq. in.?

8. In the same way find the radii of circles of areas as follows: 616 sq. ft., 1386 sq. in., 2464 sq. ft.

9. In the same way find the diameters of circles of areas as follows: 5544 sq. in., 3850 sq. ft., 49,896 sq. in.

10. What is that number which multiplied by the number just below it and also by the next greater number equals 399 times itself?

Solve the equations in Exs. 11–19:

11. $x = \frac{600}{x - 5} - 5.$ **12.** $x = \frac{680}{x + 7} + 7.$

13. 1% of $(x - 9) = \frac{82}{x + 9}.$ **14.** $\frac{x - 11}{100} = \frac{78}{x + 11}.$

15. $\frac{x + 12}{4761} = \frac{1}{x + 12}.$ **16.** $\frac{x}{10} + 1 = \frac{1200}{x - 10}.$

17. $\frac{x}{8} - 1\frac{1}{4} = 1000 \div (x + 10).$

18. $\frac{1}{10}(x + 51)(x - 51) = 4624.$

19. $\frac{1}{2}$ of $(x - 1) = 1860 \div (x + 1).$

20. What number equals 14,641 times its own reciprocal?

21. What number has the same ratio to 3721 that 1 has to 5041 times the number?

22. What is that number which, increased by 1, equals the ratio of 960 to 1 less than the number?

23. A field is 3 times as long as it is wide, and its area is 2883 sq. rd. What are its dimensions?

24. A triangle whose base equals its height has an area of 10,082 sq. in. What is the length of the base?

25. A triangle whose base equals twice its height has an area of 5041 sq. in. Required the base and height.

26. A rectangular field is 4 times as long as it is wide, and its area is 2500 sq. rd. What are its dimensions?

27. The base of a certain triangle is 4% more than the altitude. The area is 74.88 sq. in. Required the base and altitude.

28. The surface of a sphere is $12\frac{4}{7}$ times the square of its radius. What is the radius of a sphere whose surface has an area of 616 sq. in.?

CHAPTER III

FRACTIONS CONTINUED. ROOTS. SIMULTANEOUS EQUATIONS. THE COMPLETE QUADRATIC

FRACTIONS

170. Addition of fractions. — The operation of adding and subtracting fractions, as already studied (§§ 83–84, 128–129), may now be extended.

Required to add $\frac{x-y}{x+y}$ and $\frac{2xy}{x^2-y^2}$.

Reducing to the l.c.d., we have

$$\frac{x^2-2xy+y^2}{x^2-y^2}+\frac{2xy}{x^2-y^2}=\frac{x^2+y^2}{x^2-y^2}.$$

WRITTEN EXERCISE

1. $\frac{a^2b^2}{(a+b)^2}+\frac{ab}{a^2-b^2}$.

2. $\frac{xy}{x^2-y^2}+\frac{xy}{(x-y)^2}$.

3. $\frac{2a}{4a^2-b^2}+\frac{1}{2a-b}$.

4. $\frac{pqr}{p^2q^2r^2-1}+\frac{1}{pqr+1}$.

5. $\frac{xw}{xyz+yzw}+\frac{yz}{w+x}$.

6. $\frac{xy}{x^4y^4-a^2b^2}+\frac{ab}{x^2y^2+ab}$.

7. $\frac{y}{x^4+x^3y}+\frac{x}{y^4+y^3x}$.

8. $\frac{16n}{9m^2-16n^2}+\frac{4}{3m-4n}$.

9. $\frac{x^2-2xy+y^2}{x^2+2xy+y^2}+\frac{x-y}{x+y}$.

10. $\frac{a}{a^2-b^2}+\frac{b}{a^2-2ab+b^2}$.

11. $\frac{1}{x+y}+\frac{2}{x-y}+\frac{3}{x^2-y^2}$.

12. $\frac{xyz}{ax+ay+az}+\frac{1}{bx+by+bz}$.

171. Subtraction of fractions. — From $\frac{x-y}{x^2+2xy+y^2}$ to subtract $\frac{xy}{x^2-y^2}$.

1. $\frac{x-y}{x^2+2xy+y^2} = \frac{x-y}{(x+y)^2}$.

2. $\frac{xy}{x^2-y^2} = \frac{xy}{(x+y)(x-y)}$.

3. The l.c.d. is evidently $(x+y)^2(x-y)$, and we have

$$\frac{(x-y)^2}{(x+y)^2(x-y)} - \frac{xy(x+y)}{(x+y)^2(x-y)} = \frac{x^2-2xy+y^2-x^2y-xy^2}{(x+y)^2(x-y)}.$$

The denominator may be left factored. It is usually better to factor all expressions with which we may be working, so as to cancel when possible.

WRITTEN EXERCISE

1. $\frac{a}{b-c} - \frac{b}{c-a}$.

2. $\frac{x+y}{x-y} - \frac{x-y}{x+y}$.

3. $\frac{3}{a-3} - \frac{6a}{a^2-9}$.

4. $\frac{m^4-1}{m^2-1} - \frac{m^2-1}{m+1}$.

5. $\frac{a^2-b}{a^2+b} - \frac{a^2+b}{a^2-b}$.

6. $\frac{m^2+n^2}{m^2-n^2} - \frac{m+n}{m-n}$.

7. $\frac{ab}{a^2-b^2} - \frac{ab}{a-b}$.

8. $\frac{pq+1}{p^2q^2+1} - \frac{pq-1}{1+p^2q^2}$.

9. $\frac{a+b}{a^2+b^2} - \frac{ab}{a^2-b^2}$.

10. $\frac{x^2y^2}{x^4y^4-1} - \frac{xy}{x^2y^2-1}$.

11. $\frac{(2a+b)b}{b+a} - \frac{a(a-2b)}{a-b}$.

12. $\frac{x+1}{x-1} - \frac{x^2+2x+1}{x^2-2x+1}$.

13. $\frac{pqr}{p+q+r} - \frac{qr}{p^2+pq+pr}$.

14. $\frac{26a^2b^2}{4a^2b^2-1} - \frac{13ab}{2ab+1}$.

15. $\frac{a-b}{a^2+2ab+b^2} - \frac{a+b}{a^2-2ab+b^2}$.

16. $\frac{4x+y}{6x+2y} - \frac{xy}{2y+4x}$.

172. Multiplication of fractions. — Required to multiply

$$\frac{x^2 - y^2}{4x^2 + 4xy + y^2} \text{ by } \frac{2x + y}{x - y}.$$

Factoring, indicating the multiplication, and canceling,

$$\frac{(x + y)(x - y)\cdot(2x + y)}{(2x + y)^2\ (x - y)} = \frac{x + y}{2x + y}.$$

WRITTEN EXERCISE

1. $\frac{x}{y}\cdot\frac{y^2 + y}{x^2 + x}.$

2. $\frac{m^2}{n^3}\cdot\frac{mn + n}{mn - m}.$

3. $\frac{a + b}{a - b}\cdot\frac{a^2 - b^2}{(a + b)^2}.$

4. $\frac{a^2 + ab}{a^2 + 2ba}\cdot\frac{a^2 - 4b^2}{a + b}.$

5. $\frac{9x^2 - y^2}{x^2 - 9y^2}\cdot\frac{x - 3y}{3x + y}.$

6. $\frac{2m - 1}{2m + 1}\cdot\frac{4m^2 + 4m + 1}{4(m^2 - m) + 1}.$

7. $\frac{a^2b}{c^2d}\cdot\frac{cd^2}{a^2e}\cdot\frac{ce}{ad}\cdot\frac{a}{b}.$

8. $\frac{a^4 + 2a^2b^2 + b^4}{a^2 + 2ab + b^2}\cdot\frac{a + b}{a^2 + b^2}.$

9. $\frac{m + 1}{m - 1}\cdot\frac{m^2 - 1}{m^2 + 1}\cdot\frac{m^4 - 1}{4}.$

10. $\frac{a^2b + ab + b}{a^4 - a^3 + a^2}\cdot\frac{a^3}{a^2 + a + 1}.$

11. $\frac{a^2 - b^2}{a^2 - 4b^2}\cdot\frac{a + 2b}{a + b}\cdot\frac{a - 2b}{a - b}.$

12. $\frac{a}{b + c}\cdot\frac{b - c}{b}\cdot\frac{b^2 - c^2}{a^2}\cdot\frac{ab}{b - c}.$

13. $\frac{16m^6 - 25n^4}{20m^3n^2}\cdot\frac{10m^2n^3}{4m^3 + 5n^2}.$

14. $\frac{a}{b + c}\cdot\frac{b}{c + a}\cdot\frac{c^2 + (a + b)c + ab}{a^2b^2}.$

173. Division of fractions. — Required to divide

$$\frac{a+b}{a-b} \text{ by } \frac{2ab}{a^2-b^2}.$$

Multiplying by the reciprocal of the divisor, as explained in § 90, we have

$$\frac{a+b}{a-b} \cdot \frac{(a+b)(a-b)}{2ab} = \frac{(a+b)^2}{2ab}.$$

WRITTEN EXERCISE

1. $\frac{a^2-b^2}{ab} \div \frac{a+b}{a^2b^2}.$

2. $\frac{16x^2y^2-9}{2xy} \div \frac{4xy-3}{4x^2y^2}.$

3. $\frac{a^4+2a^2b^2+b^4}{a-b} \div \frac{a^2+b^2}{a^2-b^2}.$

4. $\frac{m^2-8m+7}{m^2+8m+7} \div \frac{m-7}{m+7}.$

5. $\frac{x^2-2x+1}{x^2-2x-15} \div \frac{x-1}{x-5}.$

6. $\frac{x^2+9x+20}{x^2-9x+20} \div \frac{x+5}{x-5}.$

7. $\frac{p^2-7pq-8q^2}{p+q} \div \frac{p-8q}{p^2-q^2}.$

8. $\frac{x^2+5x+6}{x+5} \div \frac{x+3}{x^2+6x+5}.$

9. $\frac{x+y}{x-y} \cdot \frac{x-y}{x+y} \div \frac{x^2-y^2}{x^2+y^2}.$

10. $\frac{a+b+c}{a-b-c} \div \frac{(a+b+c)^2}{(a-b-c)^2}.$

11. $\frac{m^2n^2+3mn+2}{mn+5} \div \frac{mn+1}{m^2n^2-25}.$

12. $\frac{x^3+3x^2y+3xy^2+y^3}{x^2-2xy+y^2} \div \frac{x+y}{x-y}.$

13. $\frac{x^3y^3-3x^2y^2+3xy-1}{x^4-y^4} \div \frac{xy-1}{x^2+y^2}.$

14 $\left(\frac{a^2+b^2}{a^2-b^2}+\frac{a+b}{a-b}\right) \div \left(\frac{a-b}{a+b}+\frac{a+b}{a-b}\right).$

FRACTIONAL EQUATIONS

174. Illustrative problem. — Solve the equation $\frac{1}{x-3} + \frac{3}{x+1} = \frac{12}{x^2-2x-3}$.

1. Multiplying both members by $(x-3)(x+1)$, we have

$$x + 1 + 3(x-3) = 12, \text{ or}$$

2. $$4x - 8 = 12.$$

3. Therefore $$4x = 20.$$

4. Therefore $$x = 5.$$

Check. Substituting 5 for x, $\frac{1}{2} + \frac{3}{6} = \frac{12}{12}$.

WRITTEN EXERCISE

Solve the following:

1. $\frac{1}{x-1} + \frac{2}{x} = \frac{4}{x^2-x}$.

2. $\frac{7}{x} + \frac{2}{x+8} = \frac{7}{8x+x^2}$.

3. $\frac{1}{x} + \frac{1}{x+1} = \frac{9}{x(x+1)}$.

4. $\frac{20}{x-1} - \frac{11}{x+1} = \frac{110}{x^2-1}$.

5. $\frac{22}{x+1} - \frac{9}{x-1} = \frac{99}{x^2-1}$.

6. $\frac{3}{2x+1} + \frac{1}{2x-1} = \frac{6}{4x^2-1}$.

7. $\frac{3}{x+1} + \frac{2}{x-5} = \frac{36}{x^2-4x-5}$.

8. $\frac{3}{x-5} - \frac{20}{x+2} = \frac{-30}{x^2-3x-10}$.

9. $3x - \frac{1}{2}(x - 1\frac{1}{2}) = 9 - \frac{1}{4}(5x - 7)$.

10. $\dfrac{20}{x-1}+\dfrac{21}{x}=\dfrac{840}{x^2-x}.$

11. $\dfrac{2x-1}{2x-2}-\dfrac{2x+1}{2x+2}=0.$

12. $x-\dfrac{2-x}{3}=4-\dfrac{1-x}{2}.$

13. $\dfrac{12}{x+2}+\dfrac{48}{4-x^2}=\dfrac{4}{2-x}.$

14. $\dfrac{3-4x}{3-3x}-\dfrac{1}{2(x-1)}=\dfrac{1}{3}.$

15. $\dfrac{2}{x+2}+\dfrac{3}{x-2}=\dfrac{4x+7}{x^2-4}.$

16. $\dfrac{x}{4}-\dfrac{x+4}{6}=\dfrac{4}{3}+\dfrac{24-x}{12}.$

17. $\dfrac{6x-3}{3x-8}+\dfrac{1-x}{x-4}-1=0.$

18. $\dfrac{3x}{2}+\dfrac{x-1}{3x}=2\left(x-\dfrac{x}{4}\right).$

19. $\dfrac{2x-5}{3}+x=3-\dfrac{2-3x}{5}.$

20. $\dfrac{2x+a}{3x-3a}+\dfrac{3x-a}{2x+2a}=\dfrac{13}{6}.$

21. $x-\dfrac{x-2}{3}+\dfrac{10+x}{5}=\dfrac{x+23}{4}.$

22. $\dfrac{2}{5x+2(x-1)}-\dfrac{1}{19}=\dfrac{1}{7x-2}.$

23. $\dfrac{x+3}{4}+\dfrac{3-x}{5}+2+\dfrac{5-x}{2}=0.$

24. $\dfrac{9}{x^2-2x+1}-\dfrac{9}{x^2+2x+1}=\dfrac{72}{(x^2-1)^2}.$

25. $\dfrac{x+1}{x+2}+\dfrac{x-1}{4-x}=0.$

26. $\dfrac{\dfrac{x}{5}+\dfrac{1}{2}}{3}-\dfrac{x-\dfrac{x}{2}}{2}+\dfrac{x}{5}=0.$

27. $\frac{3}{7}x+\frac{1}{8}(x-1)=x-4.$

28. $\dfrac{x+3}{x-2}+\dfrac{5x-2}{3(x-2)}=\dfrac{21}{5}.$

29. $\dfrac{1}{x-2}-\dfrac{2}{3-x}+2\frac{1}{2}=\frac{5}{3}.$

30. $\dfrac{x+2}{10}-\dfrac{x-\frac{1}{2}}{5}=\dfrac{5+2x}{2}.$

31. $\dfrac{2x}{a-2b}-3-\dfrac{x}{2a-b}=0.$

32. $\dfrac{8x+5}{14}-\dfrac{3-7x}{6x+2}=\dfrac{4x+6}{7}.$

33. $\dfrac{1+7x}{x-1}-\dfrac{35x+4}{2+9x}-\dfrac{28}{9}=0.$

34. $\dfrac{2x+1}{3}+\dfrac{2(1+2x)}{3(3-x)}=\dfrac{2x-5}{3}.$

35. $\dfrac{17-3x}{5}-\dfrac{2(1+2x)}{3}+\dfrac{4x-1}{3}=0.32.$

36. $\dfrac{x}{6}-\dfrac{5}{3}-\dfrac{43}{5}-2\left(\dfrac{3x}{5}-1\right)-\dfrac{1}{3}(x+8)=0.$

37. A third of the sum of a certain number and 5, plus a fifth of the difference found by taking 5 from the number, equals 6. What is the number?

38. If 1 less than a certain number is divided by 1 more than the number, the quotient is the number divided by 1 less than itself. What is the number?

SIMULTANEOUS EQUATIONS

ORAL EXERCISE

1. Given the two equations $x + 3y = 7$, $x + y = 3$; subtract member for member, the second from the first. What is the value of y in the result?

2. Given the equations $x + 7y = 10$, $x + 4y = 1$; how may we find an equation containing only y? Find it and solve for y.

3. Given the equations $x + 9y = 19$, $x + 7y = 13$; what is the value of y? How will you now find the value of x?

175. Simultaneous equations. — Equations which have the same values for the unknown quantities are called *simultaneous equations.*

For example, $x + 2y = 12$, $x + y = 7$, give by subtraction the equation $y = 5$. But if $y = 5$, we may put 5 in place of y in either equation and find the value of x. If we do this, we find that $x = 2$.

176. Illustrative problems. — 1. Find two numbers such that the sum of the first and 10 times the second is 21, and the sum of the first and 3 times the second is 7.

1. Let x = the first number and y = the second.

2. Then
$$x + 10y = 21,$$
$$x + 3y = 7.$$

3. Therefore
$$7y = 14, \text{ by subtracting,}$$
and
$$y = 2.$$

4. Substituting this value of y in the first equation,
$$x + 20 = 21, \text{ and } x = 1.$$

Check. Substitute both values in the second equation, and
$$1 + 6 = 7$$

2. Solve the equations $2x + 7y = 39$, $x + y = 7$.

If we subtract at once, the x will not disappear. Hence we multiply both members of the second equation by 2. (We might have multiplied by 7, the y disappearing by subtracting.)

1. $2x + 7y = 39$,
   $2x + 2y = 14$, the second multiplied by 2.
2. $5y = 25$, subtracting.
3. $y = 5$.
4. $x + 5 = 7$, substituting 5 for y in the second.
5. $x = 2$.

Check. Substituting in the first equation, $2\ 2 + 7\ 5 = 39$, for $4 + 35 = 39$.

WRITTEN EXERCISE

Solve the following:

1. $x + 2y = 7$,
   $x + y = 5$.

2. $x + 5y = 3$,
   $x - y = 0$.

3. $x + 6y = 1$,
   $x - 2y = 9$.

4. $x - y = 3$,
   $x + 5y = 15$.

5. $x - 7y = 3$,
   $x + 2y = 12$.

6. $x + 4y = 24$,
   $x + 2y = 14$.

7. $3x + 4y = 7$,
   $x - y = 0$.

8. $3x + 2y = 20$,
   $x - 2y = 4$.

9. $5x + y = 51$,
   $x + 3y = 13$.

10. $4x - y = 31$,
    $x + 7y = 15$.

11. $3x + y = 40$,
    $x - 2y = 4$.

12. $2x - y = 35$,
    $x + 3y = 35$.

13. $5x + 2y = 68$,
    $x + 3y = 37$.

14. $7x + 6y = 57$,
    $x - 9y = 18$.

15. $14x - 29y = 1$,
    $x - 7y = 5$.

16. $30x + 4y = -34$,
    $x - 17y = 16$.

177. Illustrative problem. — Solve the equations

$$3x - 2y = 24, \quad 2x + 7y = 41.$$

Since the coefficients of x are 3 and 2, we may make them alike by multiplying both members of the first equation by 2, and both members of the second by 3. Then

1. $6x - 4y = 48,$
2. $6x + 21y = 123.$
3. $25y = 75$, subtracting (1) from (2).
4. $y = 3$, dividing by 25.
5. Hence $3x - 6 = 24$, by substituting in the first,

and $3x = 30,\ x = 10.$

Check. Substituting in the other equation, $20 + 21 = 41$.

WRITTEN EXERCISE

Solve the following:

1. $4x + 3y = 17,$
   $3x + 7y = 27.$

2. $5x - 2y = 8,$
   $3x + 5y = 42.$

3. $5x - 3y = 8,$
   $3x + y = 30.$

4. $7x - 5y = 46,$
   $5x + 9y = 58.$

5. $4x - 3y = 33,$
   $7x - 9y = 54.$

6. $8x + 56y = 0,$
   $7x - y = 50.$

7. $9x + 2y = 66,$
   $8x - 11y = 97.$

8. $8x + 11y = -98,$
   $3x - 2y = 0.$

9. $11x - 8y = 1,$
   $7x + 5y = -70.$

10. $11x + 8y = 190,$
    $8x + 11y = 190.$

11. $\frac{1}{2}x - 3y = 33,$
    $\frac{1}{3}x + \frac{1}{9}y = 3.$

12. $\frac{1}{3}x + \frac{1}{4}y = 0,$
    $\frac{1}{2}x - \frac{1}{8}y = 4.$

13. $\frac{1}{3}y - 2x = 20,$
    $\frac{1}{3}x - 2y = -15.$

14. $\frac{1}{3}y - 7 = x,$
    $\frac{1}{3}x + 29 = y.$

15. $\frac{1}{8}x + \frac{1}{4}y = 8,$
    $\frac{1}{8}y + \frac{1}{4}x = 10.$

16. $1\%x + 2\%y = 2,$
    $2\%x + 1\%y = 2\frac{1}{8}.$

178. Elimination. — To cause one of the unknown quantities to disappear in the treatment of simultaneous equations is called *elimination* of the quantity.

For example, if $x + y = 7$, $x - y = 3$, then, by adding, $2x = 10$, and $x = 5$. Here we have *eliminated* y.

179. Methods of elimination. — Not only may an unknown quantity be eliminated by subtracting, as in §§ 176, 177, but it may be eliminated by adding, as in § 178, above. We are at liberty to eliminate the x first if we wish, or the y first, as in § 178. If the signs of the term containing the letter to be eliminated are alike, we naturally subtract; if unlike, as in § 178, we add.

It is sometimes better to eliminate by substituting at once, as in the case of $x + 3y = 17$, $3x - y = 1$.

Here we see from the first equation that $x = 17 - 3y$. Substituting this in the second, we have

1. $3(17 - 3y) - y = 1$,

or $51 - 9y - y = 1$

2. Hence $-10y = -50$, $y = 5$.
3. Therefore $x = 17 - 3y = 17 - 15 = 2$.

WRITTEN EXERCISE

Solve the following:

1. $x = 3y - 25$,
   $2x + 5y = 71$.
2. $6x - y = 51$,
   $8x + y = 47$.
3. $x + 7y = 39$,
   $5x + 2y = 30$.
4. $x = 7y - 19$,
   $5x + 3y = 19$.
5. $2x - y = -7$,
   $4x + y = -59$.
6. $7x + 4y = 15$,
   $4x - 7y = 55$.
7. $2x + 3y = 50$,
   $3x - y = 20$.
8. $x - 3y = 116$,
   $5x + y = 100$.

180. Illustrative problem. — If to the first of two numbers I add 8 times the second, the sum is 21. If from the first number I subtract twice the second, the difference is 1. What are the numbers?

1. If x = the first number and y = the second,

$$x + 8y = 21,$$
$$x - 2y = 1.$$

2. Subtracting,

$$10y = 20,$$
$$y = 2.$$

3. Substituting,

$$x - 4 = 1,$$
$$x = 5.$$

Check. Substituting in the *first equation*, because the second one has been used in finding the value of x,

$$5 + 16 = 21.$$

9. Find two numbers whose sum is 99 and whose difference is 17.

10. Find two numbers such that 7 times the first equals 8 times the second, and such that their sum is $7\frac{1}{2}$.

11. The sum of two numbers is 20, and if 14 be subtracted from the first, the result is the second. What are the numbers?

12. What is that fraction which equals $\frac{2}{3}$ when 5 is added to both terms, but equals $\frac{1}{5}$ when 2 is subtracted from both terms?

13. What is that fraction which equals $\frac{14}{15}$ when 7 is added to both terms, but equals 50% when 6 is subtracted from both terms?

14. The sum of the faces of two promissory notes is \$1000. The first draws 6% interest and the second draws 5%. The sum of the interest on each for one year is \$56. Required the face of each note.

15. $\frac{1}{5}x + \frac{1}{6}y = 4,$
$\frac{1}{2}x + \frac{1}{4}y = 8.$

16. $\frac{1}{7}x + \frac{1}{8}y = 4,$
$\frac{1}{2}x - \frac{1}{4}y = 3.$

17. $2x + 3y = 22,$
$3x + 4y = 29.$

18. $\frac{5}{8}x + \frac{3}{5}y = 31,$
$\frac{3}{8}x + \frac{1}{5}y = 17.$

19. $\frac{3}{7}x + \frac{2}{3}y = 55,$
$3x - y = 198.$

20. $\frac{1}{9}x - \frac{1}{8}y = 1,$
$\frac{1}{6}x + \frac{1}{4}y = 12.$

21. $\frac{1}{2}x + \frac{1}{3}y = 13,$
$\frac{1}{5}x + \frac{1}{8}y = 5.$

22. $2x + 3y = 17,$
$5x - 2y = 14.$

23. $\frac{1}{3}x + \frac{1}{8}y = 8,$
$\frac{1}{6}x + \frac{1}{10}y = 1.$

24. $\frac{5}{6}x + \frac{5}{8}y = 0,$
$\frac{1}{3}x + \frac{1}{8}y = 5.$

25. $\frac{1}{7}x + \frac{1}{8}y = 20,$
$\frac{3}{10}x + y = 101.$

26. $\frac{1}{7}x + \frac{1}{8}y = 7,$
$\frac{8}{9}x - y = 18.$

27. $35x + 14y = 9,$
$10x - 7y = 1.$

28. $\frac{1}{5}x + \frac{1}{6}y = 12,$
$x + \frac{2}{3}y = 59.$

29. $x + y = 1,$
$8x + 16y = 13.$

30. $4x + 7y = 22,$
$16x - y = 1.$

31. $\frac{3}{5}x - 7y = 19,$
$\frac{3}{4}x - 4y = 19.$

32. $3x - \frac{5}{6}y = 13,$
$\frac{5}{6}x - 3y = -13.$

33. $\frac{4}{5}x - y = 4,$
$\frac{7}{15}x - 3y = 41.$

34. $\frac{5}{12}x - \frac{3}{8}y = 36,$
$\frac{7}{12}x + 16 = 100.$

35. $18x - 10y = 9,$
$27x + 40y = 52.$

36. $9x - 4y = 44,$
$7x + 11y = 133.$

37. $10x + y = 7,$
$12x - 5y = -4.$

38. $7x + 11y = 8,$
$14x + 33y = 22.$

39. $\frac{1}{5}x + \frac{3}{5}y = 51,$
$2x - 3y = -30.$

40. $4x + 5y = 4,$
$12x - 5y = -2.$

41. $7x + 9y = 41,$
$8x - 11y = -17.$

42. $2x + 83y = 34,$
$15x - 79y = 255.$

43. $42x - 60 + 3y = 2x + y + 234,\ x = y.$

181. Illustrative problem. — In preparing some medicine a druggist wishes to know how much water he must add to a quart of alcohol, which already contains 5% water, so that the mixture may contain 50% alcohol.

1. Let x = the *number* of quarts of water to be added.

2. Then $50\%(1 + x) =$ number of quarts of alcohol in the mixture, which must be the 95% of a quart with which he started, since no alcohol has been added.

3. Then
$$50\%(1 + x) = 95\%,$$
$$1 + x = \tfrac{95}{50},$$
$$x = \tfrac{95}{50} - 1 = \tfrac{45}{50} = \tfrac{9}{10}.$$

4. Therefore he must add $\frac{9}{10}$ of a quart of water.

44. How many ounces of gold must be melted with 30 oz. of gold 16 carats fine ($\frac{16}{24}$ pure) to make an ingot 18 carats fine?

45. How many ounces of pure gold must be melted with 18 oz. of pure gold and 6 oz. of silver to make an ingot 22 carats fine?

46. How much water must be added to a quart of a solution containing 8% acid and the rest water, so that the new mixture shall contain 6% acid?

47. How many ounces of pure silver must be melted with 300 oz. of silver 800 fine (800 parts of pure silver in 1000 parts of metal) to make a bar 850 fine?

48. How many ounces of pure silver must be melted with 400 oz. of silver and 100 oz. of tin to make a bar 900 fine?

49. How much water must be added to a gallon of a solution containing 9% of a certain extract, so that the new mixture shall contain 4% of extract?

50. How much cotton-seed oil must be added to a pint of a mixture, which is $\frac{1}{4}$ cotton-seed oil and the rest olive oil, so that the new mixture shall contain $\frac{7}{8}$ cotton-seed oil?

182. Literal equations. — Equations in which some or all of the coefficients of the unknown quantities are letters are called *literal equations.*

For example, $ax + by = c$, $mx + ny = p$ are literal equations. The letters a, b, c, m, n, p in these equations are supposed to represent known quantities.

183. Illustrative problems. — **1.** Solve the equation

$$ax - 7b = c.$$

1. Adding $7b$ to both members,

$$ax = 7b + c.$$

2. $$x = \frac{7b + c}{a}. \quad \text{(Check the result.)}$$

2. Solve the equations $ax + by = c$, $bx + cy = d$.

1. Multiplying the first by b and the second by a,

$$abx + b^2y = bc,$$

$$abx + acy = ad.$$

2. $$(b^2 - ac)y = bc - ad.$$

3. $$y = \frac{bc - ad}{b^2 - ac}.$$

We may now find the value of x by substituting, or we may eliminate y instead of x. Taking the former plan, and substituting in the first,

4. $$ax + \frac{b(bc - ad)}{b^2 - ac} = c.$$

5. $$ax = c - \frac{b(bc - ad)}{b^2 - ac}$$

6. $$= \frac{b^2c - ac^2 - b^2c + abd}{b^2 - ac}$$

7. $$= \frac{a(bd - c^2)}{b^2 - ac}.$$

8. $$x = \frac{bd - c^2}{b^2 - ac}.$$

WRITTEN EXERCISE

Solve the equations in Exs. 1–16:

1. $ax + b = c$.
2. $a^2x - bc = ad^2$.
3. $abx + bcx = abc$.
4. $px - q = qx - p$.
5. $ax + b = cx + d$.
6. $mx - nx = ax + b$.
7. $mx + nx = px + q$.
8. $x - mx + nx - px = q$.
9. $x + ax + bx = cx + d$.
10. $(a+b)x+c=(b+c)x+d$.
11. $ax + y = m$, $bx - y = n$.
12. $x + ay = b$, $ax + y = c$.
13. $\frac{2x - y}{a + b} = 0$, $\frac{x + y}{a} = 3$.
14. $\frac{x + y}{ab} = c$, $\frac{x - y}{cd} = a$.
15. $\frac{1}{2}ax + 3y = a$, $\frac{2}{3}x + 7ay = b$.
16. $\frac{1}{2}x + \frac{1}{3}y = 2a$, $3x - 2y = 0$.

17. One number is a times another, and their sum is $2a$. What are the numbers?

18. The sum of a certain number and a equals the sum of another number and b. The first number plus the second equals c. Find the numbers.

19. The sum of two numbers is m and the sum of the first and n times the second is n. Find the numbers. What are the numbers if $n = 3$, $m = 2$?

20. If to a times a certain number there be added n times another number, the sum is b; the first number minus c times the second equals a. What are the numbers?

21. The sum of two numbers is s and their difference is d. What are the numbers? From the result write a rule for finding each of two numbers, given their sum and their difference.

Solve the equations in Exs. 22–40:

22. $ax - 3b = ab$.

23. $a^2x + 7ab = b^2$.

24. $abx - 3abc = bc^2$.

25. $a^2bx + abc = bc^2$.

26. $4ax - 3bc = 9bc$.

27. $x^2 + 2ab = a^2 + b^2$.

28. $a^2x^2 - b^2 = c^2 - 2bc$.

29. $x^2 - 4ab = a^2 + 4b^2$.

30. $16x^2 - m^2 = n^2 + 2mn$.

31. $6a^2b^2x + 5abc = 41abc$.

32. $ax + y = b$,
$bx + y = c$.

33. $ax - by = 3$,
$bx - ay = 5$.

34. $a^2x - b^2y = c^2$,
$ax - by = c$.

35. $abx - bcy = ac$,
$acx - aby = bc$.

36. $4ax + 3by = 6c$,
$5ax + 2by = 7c$.

37. $5abx + 2y = 7$,
$3acx + by = 5$.

38. $\frac{1}{4}x^2 + ax + a^2 = b^2 + ax$.

39. $x^2 - 12a^2b = 36a^4 + b^2$.

40. $a^2(x^2 - b^2) = c(2ab + c)$.

41. If the sum of two numbers is 19, and their difference is 7, what are the numbers?

42. If the sum of two numbers is 251, and their difference is 75, what are the numbers?

43. If the sum of two numbers is 24, and one is 7 times the other, what are the numbers?

44. If I have 15 cents more in one hand than in the other, and the total amount is 75 cents, how much have I in each hand?

45. Of two numbers, a times the first plus b times the second is k, and m times the first plus n times the second is l. What are the numbers?

46. Of two numbers, the sum of twice the first and 3 times the second is 38, and the sum of 5 times the first and 6 times the second is 83. What are the numbers?

QUADRATIC EQUATIONS

184. Solution of the complete or affected quadratic. — We have already learned (§ 168) how to solve the incomplete or pure quadratic equation. We shall now consider the solution of a complete quadratic equation like

$$x^2 - 8x + 15 = 0.$$

1. Factoring $x^2 - 8x + 15$, we have $(x - 3)(x - 5) = 0$.

2. It is evident that this product cannot equal 0 unless one of its factors is 0, and that if either factor is 0 the product must be 0.

3. Therefore the equation is true if

$x - 3 = 0$, in which case $x = 3$,

or if $x - 5 = 0$, in which case $x = 5$.

Check. Substituting 3 for x in the equation, $9 - 24 + 15 = 0$. Substituting 5, $25 - 40 + 15 = 0$.

ORAL EXERCISE

Solve the following equations:

1. $(x-2)(x-3)=0$.
2. $(x-1)(x-7)=0$.
3. $(x-8)(x-10)=0$.
4. $(x-9)(x-11)=0$.
5. $(x+8)(x-11)=0$.
6. $(x+9)(x-10)=0$.
7. $(x+7)(x+12)=0$.
8. $(x+6)(x+20)=0$.
9. $x(x-1)=0$.
10. $x(x-7)=0$.
11. $x(x+3)=0$.
12. $x(x+6)=0$.
13. $x(x+9)=0$.
14. $x(x+50)=0$.
15. $(2x-1)(x-1)=0$. If $2x-1=0$, $2x=1$; then what does x equal? What is the other value of x?
16. $(3x-1)(x-2)=0$.
17. $(4x-1)(x-3)=0$.
18. $(5x-2)(x-3)=0$.
19. $(7x-3)(x-5)=0$.
20. $(3x-2)(x+6)=0$.
21. $(2x-7)(3x+2)=0$.
22. $(2x+1)(3x+2)=0$.
23. $(8x+5)(5x+8)=0$.
24. $(2x-4)(3x-9)=0$.
25. $(7x-8)(8x-7)=0$.

WRITTEN EXERCISE

Solve the equations in Exs. 1–20:

1. $x^2 + x - 2 = 0.$	2. $x^2 + x - 6 = 0.$
3. $x^2 + x - 12 = 0.$	4. $x^2 - x - 12 = 0.$
5. $x^2 + 5x + 4 = 0.$	6. $x^2 + 5x + 6 = 0.$
7. $x^2 + 3x + 2 = 0.$	8. $x^2 + 4x - 5 = 0.$
9. $x^2 + 7x + 12 = 0.$	10. $x^2 + 9x + 8 = 0.$
11. $x^2 + 4x - 12 = 0.$	12. $x^2 - 9x + 20 = 0.$
13. $x^2 - 9x + 14 = 0.$	14. $x^2 + 8x + 12 = 0.$
15. $x^2 - 10x + 9 = 0.$	16. $x^2 - 4x - 32 = 0.$
17. $x^2 - 12x + 36 = 0.$	18. $x^2 - 11x + 30 = 0.$
19. $x^2 + 12x + 27 = 0.$	20. $x^2 + 12x + 35 = 0.$

21. What number is 16% of its own reciprocal?

22. Find a number which is 6 less than its square. Are there two such numbers? Are both positive?

23. Find a number which when multiplied by 1 more than itself equals 12. What are their signs?

24. Find a number whose square increased by 35 is 12 times the number itself? What are their signs?

25. If a certain number be subtracted from 16, and the difference be multiplied by the number, the product is 55. Required the number.

26. A certain rectangle is 3 yd. longer than wide. If the width be decreased by 2 yd., and the length increased by 7 yd., the area is not changed. What are its dimensions?

27. The width of a certain rectangle is 7 ft. less than its length. If the width be decreased by 4 ft., and the length be increased by 22 ft., the area is not changed. What are its dimensions?

CPSIA information can be obtained
at www.ICGtesting.com
Printed in the USA
LVHW081515041222
734562LV00012B/752

9 780526 278251